Magnetic Microscopy of Layered Structures

Magnetic Microscopy of Layered Structures

Editor

Sneha Tripathi

scitus
academics

Magnetic Microscopy of Layered Structures

Edited by **Sneha Tripathi**

Printed in 2017

ISBN: 978-1-68117-222-4

Library of Congress Control Number: 2015936582

© 2016 by
SCITUS Academics LLC,
616, Corporate Way, Suite 2, 4766,
Valley Cottage, NY 10989

www.scitusacademics.com

Contents

vi

Preface

This book presents the important analytical technique of magnetic microscopy. This method is applied to analyze layered structures with high resolution. This book presents a number of layer-resolving magnetic imaging techniques that have evolved recently. Many exciting new developments in magnetism rely on the ability to independently control the magnetization in two or more magnetic layers in micro- or nanostructures. This in turn requires techniques with the appropriate spatial resolution and magnetic sensitivity. The book begins with an introductory overview, explains then the principles of the various techniques and gives guidance to their use. Selected examples demonstrate the specific strengths of each method. Thus the book is a valuable resource for all scientists and practitioners investigating and applying magnetic layered structures.

Editor

Size-Controlled Synthesis of Fe$_3$O$_4$ Magnetic Nanoparticles in the Layers of Montmorillonite

Katayoon Kalantari, Mansor B. Ahmad, Kamyar Shameli, Mohd Zobir Bin Hussein, Roshanak Khandanlou, and Hajar Khanehzaei

Department of Chemistry, Faculty of Science, Universiti Putra Malaysia (UPM), 43400 Serdang, Selangor, Malaysia

ABSTRACT

Iron oxide nanoparticles (Fe$_3$O$_4$-NPs) were synthesized using chemical coprecipitation method. Fe$_3$O$_4$-NPs are located in interlamellar space and external surfaces of montmorillonite (MMT) as a solid supported at room temperature. The size of magnetite nanoparticles could be controlled by varying the amount of NaOH as reducing agent in the medium. The interlamellar space changed from 1.24nm to 2.85nm and average diameter of Fe$_3$O$_4$ nanoparticles was from 12.88 nm to 8.24nm. The synthesized nanoparticles were characterized using some instruments such as transmission electron microscopy, powder X-ray diffraction, energy dispersive X-ray spectroscopy, field emission scanning electron microscopy, vibrating sample magnetometer, and Fourier transform infrared spectroscopy.

INTRODUCTION

In recent years, considerable attention has been paid to iron oxides, especially on magnetic (Fe_3O_4) nanoparticles due to their potential applications such as pigment, magnetic resonance imaging, magnetic drug delivery, Ferro fluids, recording material, and data storage media [1, 2]. There are many reports on the synthesis and characterization of magnetic nanoparticles; this is due to its significance in various fields, especially, when this material is made at the nanoscale. Magnetic nanoparticles have been prepared by various methods such as arc discharge, mechanical grinding, laser ablation, microemulsion, and high temperature decomposition of organic precursors [3]. However, research is still going on to obtain well-dispersed magnetic nanoparticles.

Several methods have been developed recently for preparing magnetic nanoparticles, including coprecipitation [4], sol-gel method [5], hydrothermal process [6], and the solvothermal method [7]. Among these methods, coprecipitation is considered as the simplest, most efficient, and most economic method.

However, there are three major challenges that these nanoparticles present. One is related to easy oxidation/dissolution of magnetic nanoparticles (NPs). The other two drawbacks are the difficulty in recycling such small sized NPs and coaggregation of nanoparticles in which case, leads to decrease the effective surface area of nanoparticles and thus reduce their reaction activities. In order to protect the magnetic NPs against oxidation, a shell structure is often introduced, such as a silica shell or polymer shell [8]. Several methods have been accordingly developed to minimize the coaggregation of nanoparticles, obtain the soft sediment, and improve their manipulation, such as supporting of magnetic nanoparticles on polymers or inorganic matter, like porous silica [9, 10].

Clay minerals, which are abundant and environmental friendly, are used as the supporting materials of magnetic nanoparticles. Sodium montmorillonite (Na-MMT) ($Na_{0.7}(Al_{3.3}Mg_{0.7})$ $Si_8O_{20}(OH)_4 \cdot nH_2O$), a kind of 2:1 type layered silicates in which each layer comprises an alumina octahedral sheet sandwiched between two silica tetrahedral sheets, is known as a polymer modifier due to its high specific surface area [11].

Such layers are stacked by weak dipolar or van der Waals forces, leading to the interaction of charge compensating cations into the interlayer space. Therefore, not only adsorption on the external surface but also intercalation into the interlayer space can occur [12].

Here we present our results obtained from synthesized magnetic nanoparticles well dispersed on montmorillonite using sodium hydroxide as a reducing agent.

MATERIALS AND METHODS

In this work, all reagents were of analytical grade. Ferric chloride hexahydrate (FeCl$_3$.6H$_2$O) and ferrous chloride tetrahydrate (FeCl$_2$.4H$_2$O) of 96 wt. % were used as the iron precursors and also montmorillonite powder was obtained from (Kunipia-F, Tokyo, Japan). NaOH of 99 wt. % was obtained from Merck KGaA (Darmstadt, Germany). All solutions were freshly prepared using deionized water.

Fe$_3$O$_4$/Montmorillonite NCs Preparation

The Fe$_3$O$_4$ nanoparticles were synthesized using chemical coprecipitation method. Prior to use, all glassware used in experimental procedures were cleaned in a fresh solution of HNO$_3$/HCl (3:1, v/v), washed thoroughly with double distilled water, and dried. For the synthesis of Fe$_3$O$_4$/montmorillonite nanocomposites (Fe$_3$O$_4$/MMT NCs), different volumes of deionized water were bubbled by N$_2$ gas for 15 min and then 2.0g of montmorillonite and measured amount of Fe^{3+} and Fe^{2+} with a molar ratio of 1:2 were successively dissolved in ultrapure water with vigorous mechanical stirring for 2 hours. Under the protection of nitrogen gas, 1.50, 3.0, 5.0, 7.50, 9.50, and 12.50 mL of 2M NaOH were added dropwise into the above solutions. Eventually the Fe$_3$O$_4$-NPs content of the MMT matrix was 1.0, 3.0, 5.0, 7.0, 9.0, and 12.0 wt. %, respectively. After that, the mechanical stirrer was switched off and Fe$_3$O$_4$-NPs settled gradually. The suspensions were finally centrifuged, washed twice with ethanol and distilled water, and kept in a vacuum stove at 100°C.

Characterization

The structures of the synthesized Fe_3O_4-NPs in MMT were examined using Philips X'pert PXRD (copper $K\alpha$ radiation; PANalytical, Almelo, The Netherlands). The changes in the interlamellar spacing of MMT were also studied by using PXRD in the small-angle range of 2θ (5–15 degrees). The scan speed of 2 degrees/minutes was applied to PXRD patterns recording. TEM images of the samples were recorded with an H-7100 electron microscope (Hitachi Ltd., Tokyo, Japan). The samples were prepared by dispersing about 0.1 g powder with 8 mL ethanol by ultrasound for 20 min. A drop of the dispersion was applied to a holey carbon TEM support grid. Excess solution was blotted off using a filter paper. The analysis of the surface morphologies of the samples was performed with an (XL 30; Philips) environmental scanning electron microscope (SEM). The cleaned and dried samples were first coated with gold using sputter coater. For elemental analysis of the nanoparticles, energy dispersion X-ray spectroscopy was carried out on a Shimadzu EDX-700HS spectrometer attached to the SEM. The FTIR spectrum was used to identify the functional groups present in the synthesized compound. FTIR spectra were recorded over the range of 200–4000 cm^{-1} utilizing the Series 100 FTIR 1650 spectrophotometer (PerkinElmer, Waltham, MA, USA). Magnetization measurements were carried out with a Lakeshore (model 7407) vibrating sample magnetometer (VSM) to study magnetic properties of the Fe_3O_4 nanoparticles under magnetic field up to 10 kG at room temperature.

RESULTS AND DISCUSSION

The customized shape and size of prepared magnetic nanoparticles have been a challenge in scientific and technological issues. The wet chemical methods to magnetic nanoparticles are more effective and simpler with good control over composition and size [13]. Fe_3O_4-NPs can be prepared through the coprecipitation of Fe^{3+} and Fe^{2+} aqueous salts solution by addition of base as reducing agent. The chemical reaction may be presented as follows [14]:

$$Fe^{2+}/Fe^{3+}/MMT + 8OH^- \longrightarrow Fe_3O_4/MMT\ NCs + 4H_2O. \quad (1)$$

The surface of MMT is assisting the Fe$_3$O$_4$-NPs nucleation during the reduction process. The schematic illustration of the synthesis of Fe$_3$O$_4$/MMT NCs from MMT/Fe^{3+}–Fe^{2+} suspension produced by using sodium hydroxide is shown in Figure 1. Meanwhile, as shown in Figure 2, the MMT suspension was gray (a) and the addition of Fe^{3+}–Fe^{2+} ions to the MMT change color to the light brown (b), brown (c), and dark brown (d) for 1.0, 7.0, and 12.0 wt. %, respectively, but after the addition of NaOH to the suspensions they turned to black color (e), (f), and (g) that indicate the abundance of Fe$_3$O$_4$-NPs in the MMT suspensions.

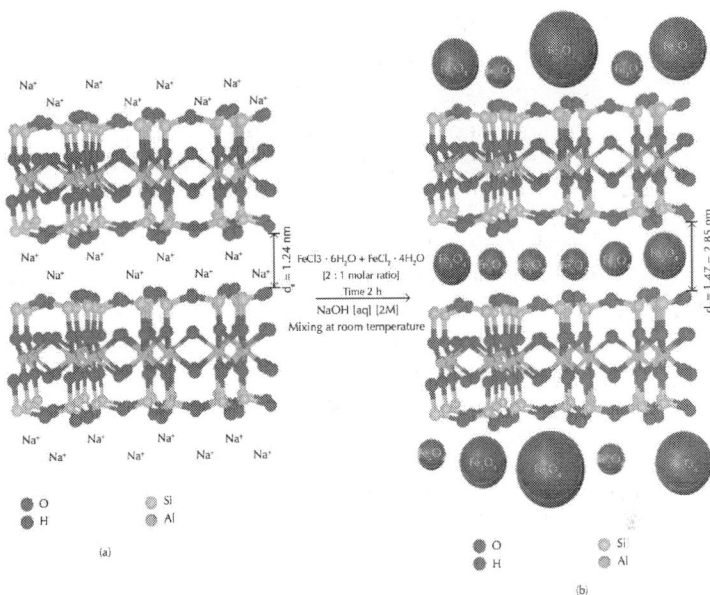

Figure 1: Schematic illustration of the synthesized Fe$_3$O$_4$-NPs in the interlayer space of MMT suspension by chemical coprecipitation reduction method ((a)-(b)).

(a)

(b)

(c)

(d)

(e)

(f)

(g)

Figure 2: Photograph of montmorillonite suspension (a) and Fe^{2+}/Fe^{3+}/MMT suspension at different Fe^{2+}/Fe^{3+}/concentrations; (b) 1.0%, (c) 7.0%, (d) 12.0% and Fe$_3$O$_4$/MMT NCs, (e) 1.0%, (f) 7.0%, and (g) 12.0%.

Powder X-Ray Diffraction

The basal spacing of MMT (Figure 3(a)) was 1.24 nm at 2θ°, 8.83°. Combination with Fe$_3$O$_4$ NPs expanded the basal spacing to 1.47, 1.68, 1.87, 2.10, 2.53, and 2.85 nm at the 2θ°, 8.75°, 8.66°, 8.51°, 8.21°, 7.72°, and 7.46° in 1.0, 3.0, 5.0, 7.0, 9.0, and 12.0 wt. %, respectively. It may be assumed that the iron ions penetrated into the interlayer space of montmorillonite via ion-exchange and were reduced by NaOH to Fe$_3$O$_4$ NPs. In this case, the interlayer space would have acted as a size controller and microreactor [15]. Moreover the intensities of the reflections were significantly decreased, and the highly ordered parallel lamellar structure of the MMT was disrupted by the Fe$_3$O$_4$-NPs formation [16]. Figure 3(b) reveals X-ray diffraction patterns of prepared Fe$_3$O$_4$-NPs. The reflections of Fe$_3$O$_4$-NPs at 2θ°, 31.62°, 35.78°, 43.84°, 59.67°, 62.53°, and 74.58° are related to the 220, 311, 400, 511, 440, and 622 crystalline structure with a spinel structure. Comparing XRD pattern of synthesized NPs with the standard diffraction spectrum (ref. code Fe_3O_4:01-088-0315), the synthesized product is crystalline Fe_3O_4 [17]. The size of Fe_3O_4-NPs is significantly influenced by the kind of reducing agent used in the reaction. Generally, a strong reducing agent promotes a fast rate and favors the formation of smaller NPs. Meanwhile, we have found that the amount of NaOH has a significant effect on the size of Fe_3O_4-NPs [18] and increasing the NaOH amount will lead up to the decrease of Fe_3O_4-NPs size gradually. The possible reason is that repulsive force between hydroxide ions hinders the growth of crystal grains when the excess hydroxide ions produced from NaOH are adsorbed on the surface of crystal nuclei [19]. Additionally, as shown in Figure 3(b), Fe_3O_4/MMT NCs, 12.0 wt. %, completely absorbed when when exposed to an external magnetic field, and the response decreased with the decreasing in Fe_3O_4 percentage.

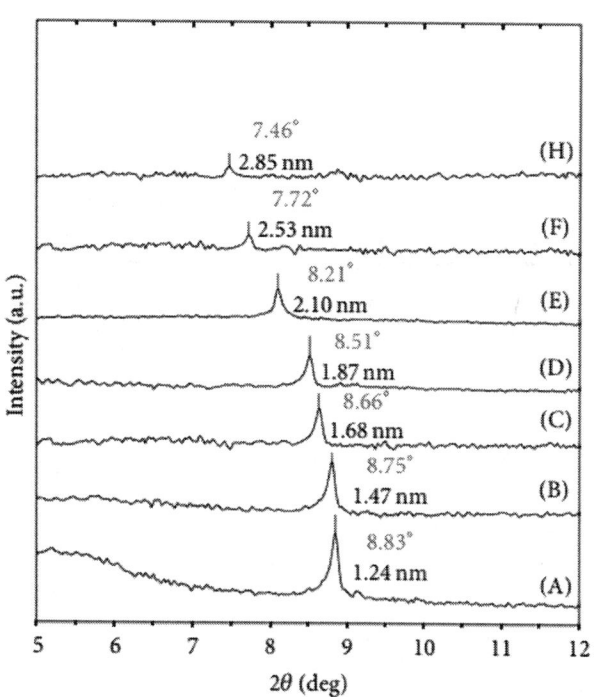

(a)

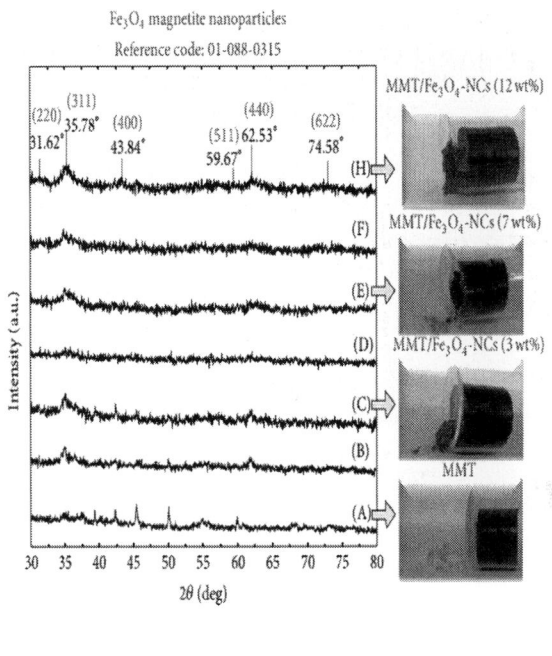

(b)

Figure 3: (a) Powder X-ray diffraction patterns of MMT (a) and Fe_3O_4/MMT NCs for determination of d-spacing (d_s) and crystals structure at different percentage of NPs (1.0, 3.0, 5.0, 7.0, 9.0, and 12.0 wt. % ((b)–(f))). (b) Powder X-ray diffraction patterns of MMT (a) and MMT–Fe_3O_4-NCs for determination of Fe_3O_4 crystal structure. (b)–(h): 1, 3, 5, 7, 9, and 12% (w/w).

Transmission Electron Microscopy (TEM)

Figure 4 shows TEM images of Fe_3O_4-NPs in the interlayer space or on the montmorillonite surface. The size distribution histograms of Fe_3O_4/MMT NCs in Figures 4(a)–4(d) were determined by measuring the diameter of 8.24, 10.38, 12.57, and 12.88 nm for 1.0, 5.0, 9.0, and 12.0 wt. %, respectively. Due to the high density of ion-exchange sites on MMT, the highly charged Fe_3O_4-NPs are strongly bound via electrostatic interaction to the surface. The higher abundance of NPs aggregates in the MMT composite is in direct correlation with the smaller primary NPs dimensions [20].

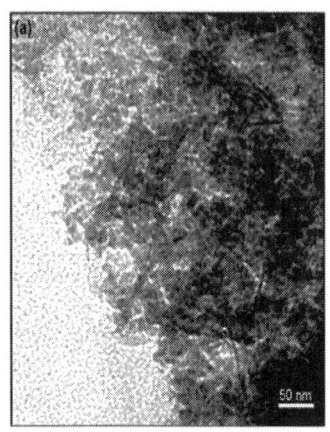
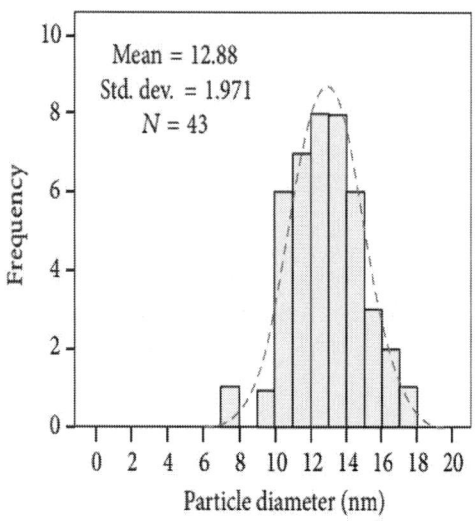

(a)

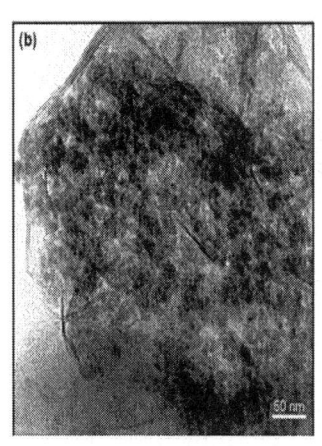
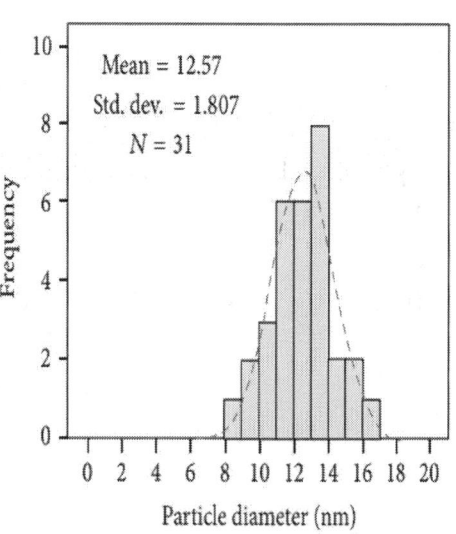

(b)

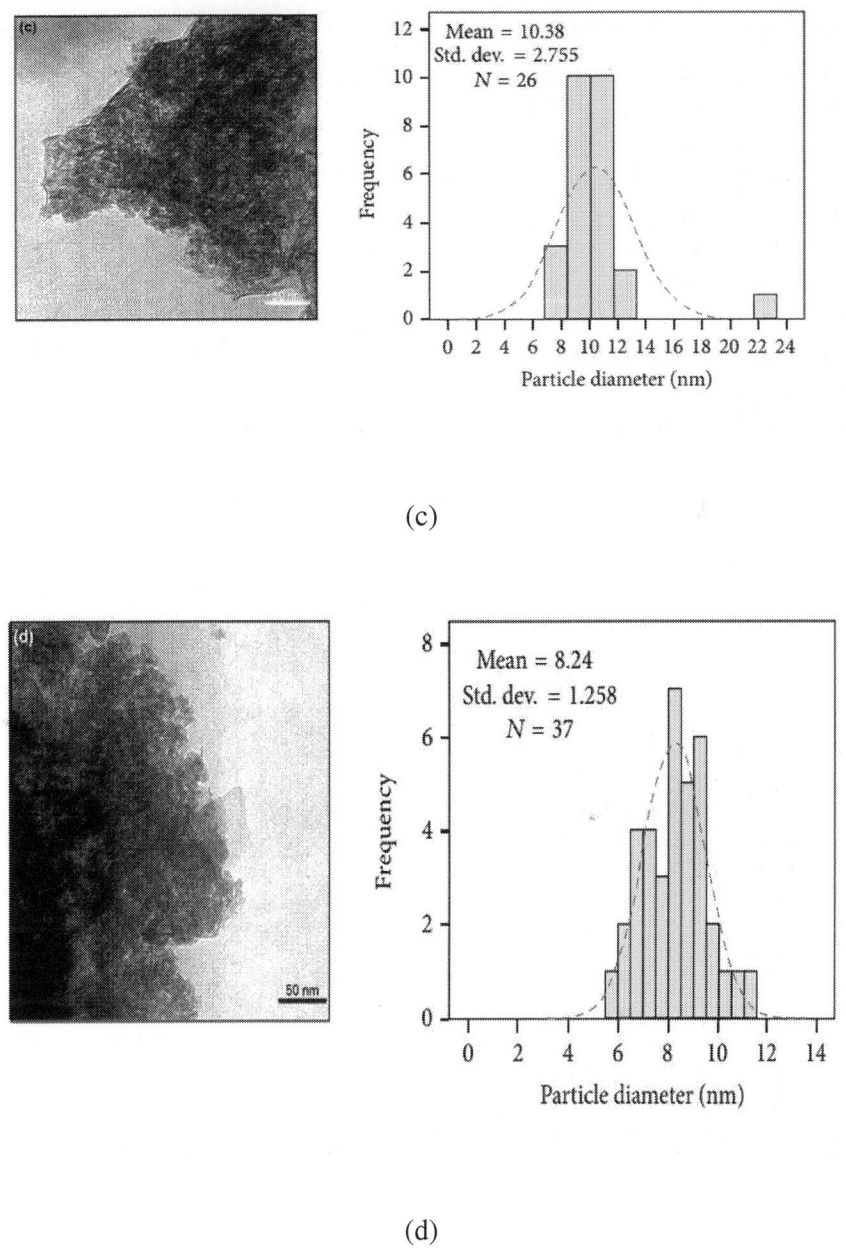

(c)

(d)

Figure 4: The transmission electron microscopy images and particle size distribution histogram for 1.0, 5.0, 9.0, and 12.0 wt. % of the Fe$_3$O$_4$/MMT NCs.

Scanning Electron Microscopy (SEM)

SEM images show that the Fe_3O_4-NPs are in spherical shape (Figures 5(a) and 5(b)). Fe_3O_4-NPs have very high surface free energy, which lead to instability in their dispersions thermodynamic system; stability is restored upon Ostwald ripening [21]. The isotopic nucleation rate per unit area at the interface between the Fe_3O_4-NPs leads to forming the spherical shape NPs due to equivalent growth rate along every direction of the nucleation as the sphere has the smallest surface area per unit volume compared to other shapes [22]. Figure 6clearly shows that the size of Fe_3O_4-NPs in 1.0 wt. % sample is smaller than 12.0 wt. % which is in line with the TEM and XRD results previously mentioned.

(a)

(b)

Figure 5: The scanning electron microscopy micrographs of Fe$_3$O$_4$/MMT NCs with high magnification for 1.0 and 12.0 wt. % (a) and (b).

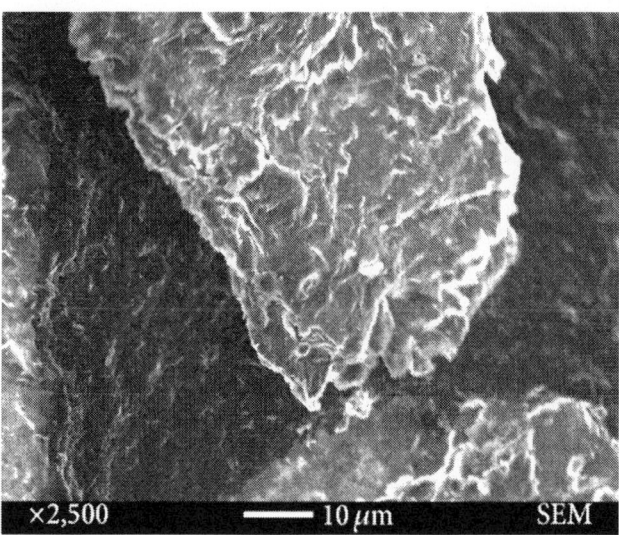

(a)

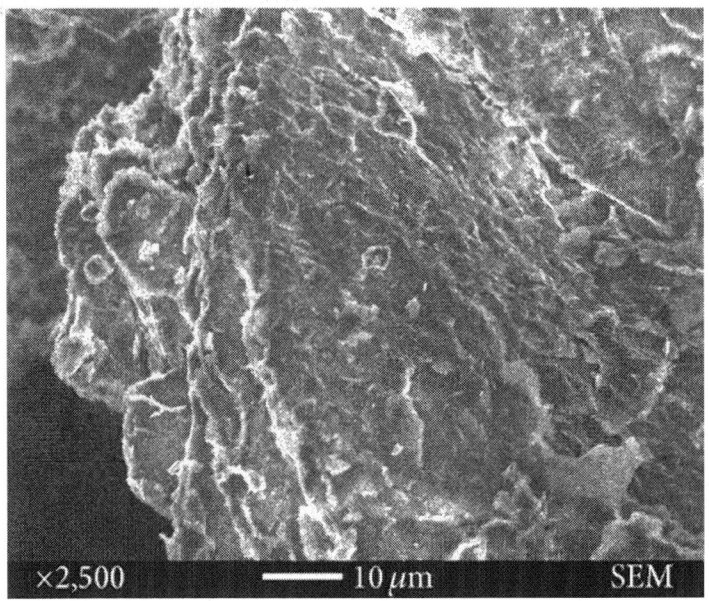

(b)

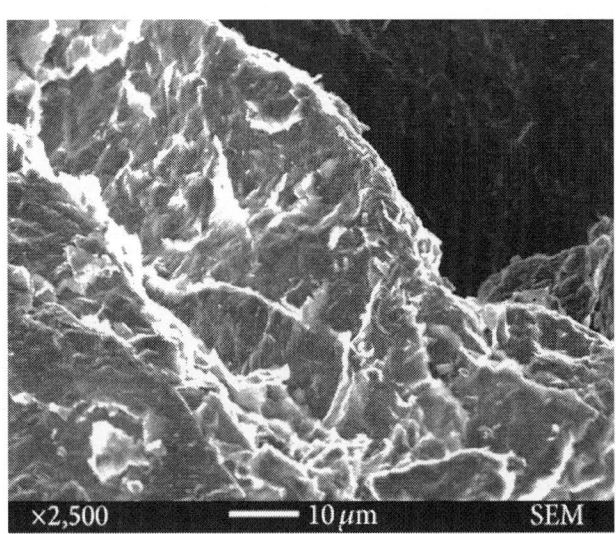

(c)

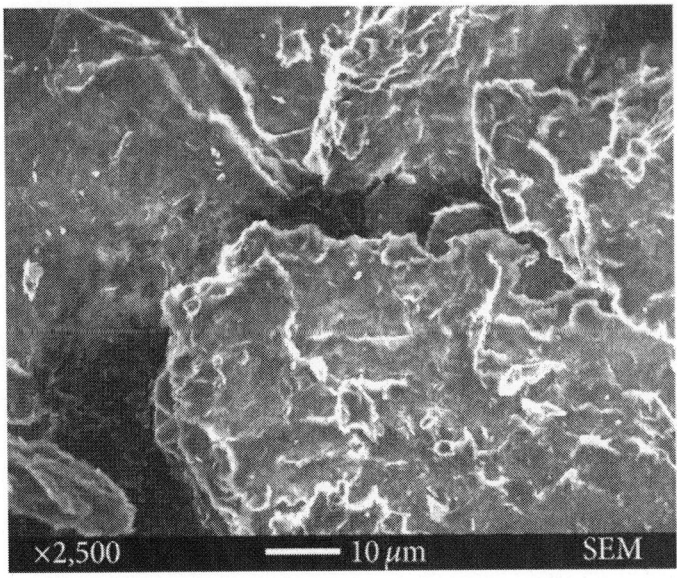

(d)

Figure 6: The scanning electron microscopy micrographs of MMT (a) and Fe₃O₄/MMT NCs for 5.0, 9.0, and 12.0 wt. % (b), (c), and (d).

A general view of the MMT with a typical sheet-like structure with large flakes can be observed in Figure 6(a) and it confirmed that the small difference can be observed in morphology structure of MMT compared with Fe₃O₄/MMT NCs (Figures 6(b)-6(c)). The different brightness of the MMT and MMT supported Fe₃O₄-NPs and some disordering in the MMT structure are due to the abundance of iron oxide coating on the surface of clay.

Energy Dispersive X-Ray Spectroscopy (EDX)

The oxygen and iron peaks in EDX spectrum reveal the existence of Fe₃O₄-NPs (Figure 7). In Figure 7(a), the peaks around 1.7, 2.7, 2.9, 3.7, 4.0, 4.5, 6.4, and 7.1 keV are related to the binding energies of MMT. The peaks around 0.2, 2.2, 6.4, and 7.0 keV are related to Fe₃O₄-NPs elements, respectively, in Figures 7(b)–7(d) [20]. Moreover, with an increase amount of iron oxide, the iron peaks height increases in the Fe_3O_4/MMT NCs (1.0, 5.0, and 12.0 wt. wt. %). Additionally, the EDXRF spectra for the MMT and for Fe_3O_4/MMT NCs confirm the presence of elemental compounds in the MMT and Fe_3O_4-NPs without any impurity peaks. The results indicate that the synthesized Fe_3O_4-NPs are in high purity.

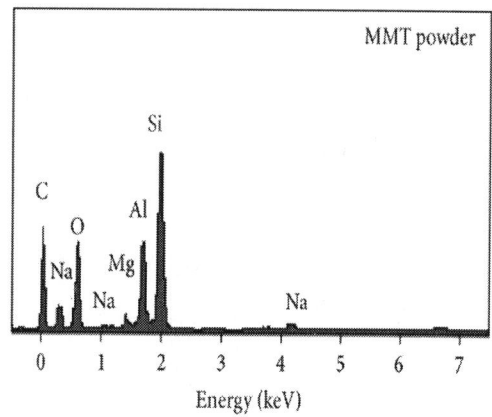

(a)

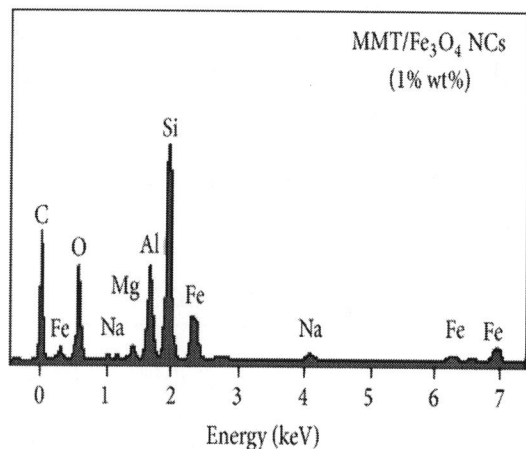

(b)

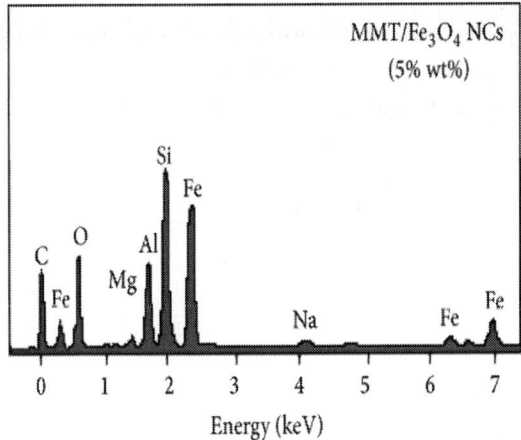

(c)

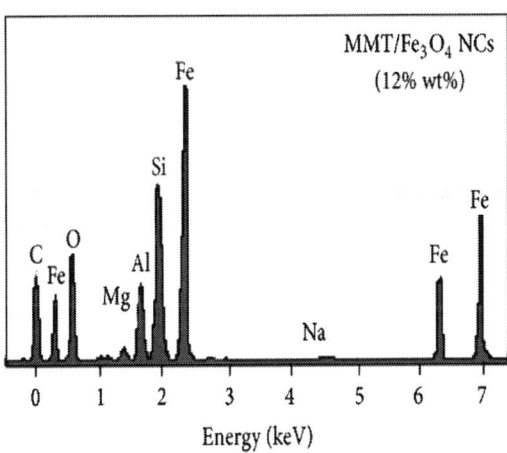

(d)

Figure 7: Energy dispersive X-ray spectroscopy of MMT (a) and Fe$_3$O$_4$/MMT NCs for 1.0, 5.0, and 12.0 wt. % (b), (c), and (d).

FTIR Chemical Analysis

For pure MMT, the bending mode of Al-Al-OH is found at $910\,cm^{-1}$ because of the large amount of Al in the octahedral site of oxygen and the bending mode of Si-O-Al was observed at $513\,cm^{-1}$. The band at $3633\,cm^{-1}$ (Figure 8(a)) that identified stretching vibration of the structural hydroxyl group of MMT was detected [23]. Figures 8(b)–8(f) show that there were no many changes in the spectra of Fe_3O_4/MMT NCs compared with MMT. The bands between $446\,cm^{-1}$ and $625\,cm^{-1}$ can be assigned to Fe-O stretching vibration which may be due to the overlapping of Si-O and Al-OH.

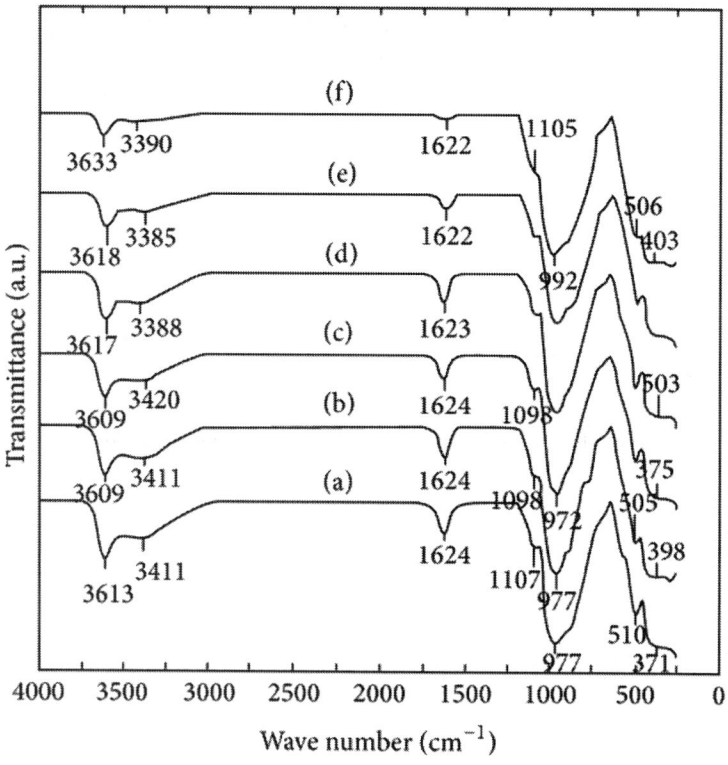

Figure 8: FTIR spectra of MMT (a) and Fe_3O_4/MMT NCs (b), (c), (d), (e), and (f).

The FTIR spectra demonstrated the inflexibility of silicate layers and nonbond chemical interface between the silicate layers and Fe$_3$O$_4$-NPs in Fe$_3$O$_4$/MMT NCs. These results confirmed that, with an increase amount of Fe$_3$O$_4$-NPs in the Fe$_3$O$_4$/MMT NCs due to the existence of van der Waals interactions between the oxygen groups of MMT and Fe$_3$O$_4$-NPs, peak areas shifted to low wave numbers and the intensity of peaks decreased [16].

Vibrating Sample Magnetometer (VSM)

The vibrating sample magnetometer (VSM) was applied to test the magnetic properties of Fe$_3$O$_4$/MMT NCs. The magnetization curves for Fe$_3$O$_4$/MMT NCs in 1.0 and 12.0 wt. % are shown in Figure 9. It can be seen from the magnetization curves that the saturation magnetization (Ms) of the Fe$_3$O$_4$/MMT NCs increased from 12.10 to 32.40 emu.g^{-1} when Fe$_3$O$_4$ content increased from 1.0 to 12.0 wt. % of composite. It indicated more Fe$_3$O$_4$ being trapped in the MMT layers.

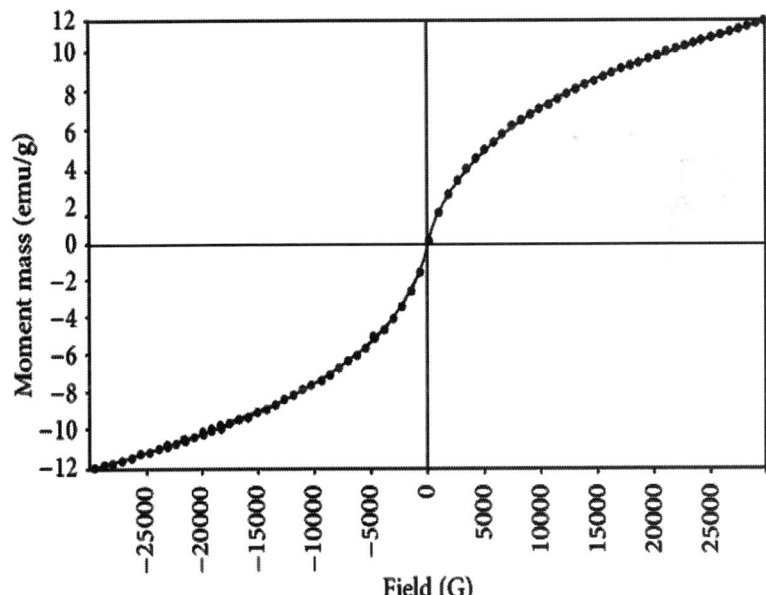

(a)

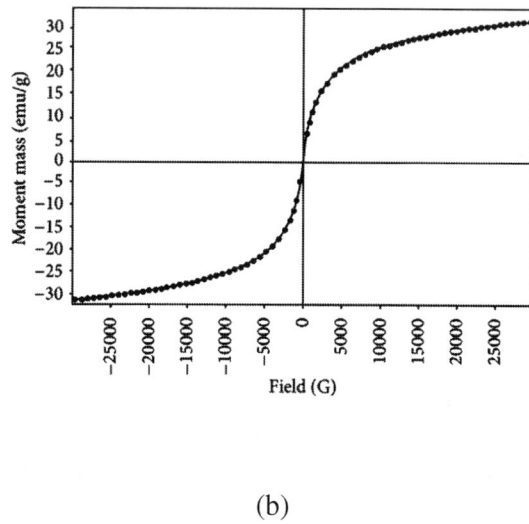

(b)

Figure 9: Magnetization curve of Fe$_3$O$_4$/MMT NCs 1.0% (a) and 12.0% (b).

CONCLUSIONS

In this work, we synthesized successfully Fe$_3$O$_4$/MMT NCs through coprecipitation method which is a cheap, facile, and environmental friendly method. Sodium hydroxide was used as reducing agent and ferric chloride and ferrous chloride were used as the iron precursors. These synthesized Fe$_3$O$_4$-NPs with high magnetization and good dispersion are on the external surface and between the layers of montmorillonite. Our results show that the amount of NaOH (reducing agent) has significant effect on the size of nanoparticles. Smaller sized nanoparticles were obtained due to more reducing agent amount. Moreover, these results indicate that montmorillonite used here can act as an effective support for the synthesis of magnetic nanoparticles.

The specific surface areas of Fe$_3$O$_4$/MMT NCs were larger than pure montmorillonite. The comparison between the PXRD patterns of MMT and the prepared Fe$_3$O$_4$/MMT NCs indicated the immobilized formation of Fe$_3$O$_4$-NPs in the interlamellar space of the MMT layers. The TEM images showed that the mean diameter of the nanoparticles ranged from 8.24 nm to 12.88 nm.

ACKNOWLEDGMENTS

The authors would like to acknowledge the financial support from Universiti Putra Malaysia (UPM) (RUGS Project no. 9199840). They are also grateful to the staff of the Department of Chemistry, UPM, and the Institute of Bioscience, UPM, for the technical assistance.

REFERENCES

1. G. Mihoc, R. Ianoş, C. Păcurariu, and I. Lazău, "Combustion synthesis of some iron oxides used as adsorbents for phenol and p-chlorophenol removal from wastewater," Journal of Thermal Analysis and Calorimetry, vol. 112, no. 1, pp. 391–397, 2013.

2. M. Abbas, B. Parvatheeswara Rao, S. M. Naga, M. Takahashi, and C. Kim, "Synthesis of high magnetization hydrophilic magnetite (Fe_3O_4) nanoparticles in single reaction—surfactantless polyol process," Ceramics International, vol. 39, no. 7, pp. 7605–7611, 2013.

3. F. Ahangaran, A. Hassanzadeh, and S. Nouri, "Surface modification of Fe_3O_4@SiO_2 microsphere by silane coupling agent," International Nano Letters, vol. 3, no. 1, pp. 1–5, 2013.

4. N. Wang, L. Zhu, D. Wang, M. Wang, Z. Lin, and H. Tang, "Sono-assisted preparation of highly-efficient peroxidase-like Fe_3O_4 magnetic nanoparticles for catalytic removal of organic pollutants with H_2O_2," Ultrasonics Sonochemistry, vol. 17, no. 3, pp. 526–533, 2010.

5. K. Raja, S. Verma, S. Karmakar, S. Kar, S. J. Das, and K. S. Bartwal, "Synthesis and characterization of magnetite nanocrystals," Crystal Research and Technology, vol. 46, no. 5, pp. 497–500, 2011.

6. S. Ahmadi, C.-H. Chia, S. Zakaria, K. Saeedfar, and N. Asim, "Synthesis of Fe_3O_4 nanocrystals using hydrothermal approach," Journal of Magnetism and Magnetic Materials, vol. 324, no. 24, pp. 4147–4150, 2012.

7. G. Gao, R. Shi, W. Qin et al., "Solvothermal synthesis and characterization of size-controlled monodisperse Fe_3O_4

nanoparticles," Journal of Materials Science, vol. 45, no. 13, pp. 3483–3489, 2010.

8.   G. Li, Z. Zhao, J. Liu, and G. Jiang, "Effective heavy metal removal from aqueous systems by thiol functionalized magnetic mesoporous silica," Journal of Hazardous Materials, vol. 192, no. 1, pp. 277–283, 2011.

9.   I. Larraza, M. López-Gónzalez, T. Corrales, and G. Marcelo, "Hybrid materials: magnetite-polyethylenimine-montmorillonite, as magnetic adsorbents for Cr(VI) water treatment," Journal of Colloid and Interface Science, vol. 385, no. 1, pp. 24–33, 2012.

10. B. C. Munoz, G. W. Adams, V. T. Ngo, and J. R. Kitchin, "Stable magnetorheological fluids," Google Patents, 2001.

11. A. K. Mishra, S. Allauddin, R. Narayan, T. M. Aminabhavi, and K. V. S. N. Raju, "Characterization of surface-modified montmorillonite nanocomposites," Ceramics International, vol. 38, no. 2, pp. 929–934, 2012.

12. M. Fan, P. Yuan, J. Zhu et al., "Core-shell structured iron nanoparticles well dispersed on montmorillonite," Journal of Magnetism and Magnetic Materials, vol. 321, no. 20, pp. 3515–3519, 2009.

13. A. K. Gupta and M. Gupta, "Synthesis and surface engineering of iron oxide nanoparticles for biomedical applications," Biomaterials, vol. 26, no. 18, pp. 3995–4021, 2005.

14. M. Zhang, G. Pan, D. Zhao, and G. He, "XAFS study of starch-stabilized magnetite nanoparticles and surface speciation of arsenate," Environmental Pollution, vol. 159, no. 12, pp. 3509–3514, 2011.

15. S. Li, P. Wu, H. Li et al., "Synthesis and characterization of organo-montmorillonite supported iron nanoparticles," Applied Clay Science, vol. 50, no. 3, pp. 330–336, 2010.

16. K. Shameli, M. B. Ahmad, M. Zargar, W. M. Z. W. Yunus, A. Rustaiyan, and N. A. Ibrahim, "Synthesis of silver nanoparticles in montmorillonite and their antibacterial behavior," International Journal of Nanomedicine, vol. 6, pp. 581–590, 2011.

17. K. Kalantari, M. B. Ahmad, K. Shameli, and R. Khandanlou, "Synthesis of talc/Fe_3O_4 magnetic nanocomposites using chemical co-precipitation method," International Journal of Nanomedicine, vol. 8, pp. 1817–1823, 2013.

18. H. Maleki, A. Simchi, M. Imani, and B. F. O. Costa, "Size-controlled synthesis of superparamagnetic iron oxide nanoparticles and their surface coating by gold for biomedical applications," Journal of Magnetism and Magnetic Materials, vol. 324, no. 23, pp. 3997–4005, 2012.

19. H. Yan, J. Zhang, C. You, Z. Song, B. Yu, and Y. Shen, "Influences of different synthesis conditions on properties of Fe$_3$O$_4$ nanoparticles," Materials Chemistry and Physics, vol. 113, no. 1, pp. 46–52, 2009.

20. T. Szabó, A. Bakandritsos, V. Tzitzios et al., "Magnetic iron oxide/clay composites: effect of the layer silicate support on the microstructure and phase formation of magnetic nanoparticles," Nanotechnology, vol. 18, no. 28, Article ID 285602, 2007.

21. S. Hribernik, M. Sfiligoj-Smole, M. Bele, S. Gyergyek, J. Jamnik, and K. Stana-Kleinschek, "Synthesis of magnetic iron oxide particles: development of an in situ coating procedure for fibrous materials," Colloids and Surfaces A: Physicochemical and Engineering Aspects, vol. 400, pp. 58–66, 2012.

22. D. K. Kim, M. Mikhaylova, Y. Zhang, and M. Muhammed, "Protective coating of superparamagnetic iron oxide nanoparticles," Chemistry of Materials, vol. 15, no. 8, pp. 1617–1627, 2003.

23. Y. H. Son, J. K. Lee, Y. Soong, D. Martello, and M. Chyu, "Structure-property correlation in iron oxide nanoparticle-clay hybrid materials," Chemistry of Materials, vol. 22, no. 7, pp. 2226–2232, 2010.

Single-Crystalline Chromium Silicide Nanowires and their Physical Properties

Han-Fu Hsu[1], Ping-Chen Tsai[1], and Kuo-Chang Lu[1, 2]

[1]Department of Materials Science and Engineering, National Cheng Kung University, No.1, University Rd, Tainan 701, Taiwan

[2]Center for Micro/Nano Science and Technology, National Cheng Kung University, No.1, University Rd, Tainan 701, Taiwan

ABSTRACT

In this work, chromium disilicide nanowires were synthesized by chemical vapor deposition (CVD) processes on Si (100) substrates with hydrous chromium chloride ($CrCl_3 \cdot 6H_2O$) as precursors. Processing parameters, including the temperature of Si (100) substrates and precursors, the gas flow rate, the heating time, and the different flow gas of reactions were varied and studied; additionally, the physical properties of the chromium disilicide nanowires were measured. It was found that single-crystal $CrSi_2$ nanowires with a unique morphology were grown at 700°C, while single-crystal Cr_5Si_3 nanowires were grown at 750°C in reducing gas atmosphere. The crystal structure and growth direction were identified, and the growth mechanism was proposed as well. This study with magnetism, photoluminescence, and field emission measurements demonstrates that $CrSi_2$ nanowires are attractive choices for future applications in magnetic storage, photovoltaic, and field emitters.

BACKGROUND

Recently, transition metal silicide nanowires have been widely studied [1]-[9] for their utilization in semiconductor device technologies. Low-resistivity silicides, such as $TiSi_2$, $CoSi_2$, and $NiSi$, have been applied for interconnection in CMOS devices [10]. The group of refractory semiconducting silicides, composed of silicon and metals, have different physical properties that are useful and importantly meaningful. Among them, semiconducting silicides, such as $CrSi_2$ and ß-$FeSi_2$, with a narrow energy gap (0.1 to 0.9 eV) have been extensively investigated for their potential use in silicon-integrated optoelectronic devices [11] such as LEDs [12], [13] and IR detectors [14]. In particular, $CrSi_2$ is a narrow bandgap (0.35 eV) semiconductor [15]-[17], offering applications in the Schottky barrier solar cell technology [18]. Hexagonal $CrSi_2$ with a C40-type structure has a high melting point and excellent resistance to oxidation, deformation, and stretching, being considered to be a potential structural material for aerospace and energy generation industries [19]. Additionally, it is a thermoelectric conversion component that could be applied to generate electric power at high temperatures [20]; the figure of merit (ZT) of $CrSi_2$ has been measured to be 0.25 at 900 K [21]. $CrSi_2$ also has good field emission with relatively low work function (3.9 eV) [22] as compared with generally studied field emission materials such as CNTs (5 eV) [23] and ZnO (5.3 eV) [24]. With excellent intrinsic properties of $CrSi_2$, one-dimensional $CrSi_2$ nanowires are expected to improve field emission performances by bulk and thin film $CrSi_2$. Though there have been some previous studies on $CrSi_2$ nanowires [25]-[28], two special aspects can be found in this research. Firstly, we conducted a more systematic study on the influences of each processing parameter on growth. Secondly, we provided a low-cost and simple method to synthesize high-quality $CrSi_2$ nanowires with very good physical properties.

METHODS

In our experiments, we synthesized chromium disilicide nanowires with chemical vapor deposition (CVD) processes. Single-crystal Si (001) wafers, the native oxide of which was etched by BOE solution, were substrates. The metal source was from hydrous chromium chloride ($CrCl_3 \cdot 6H_2O$) powders, and the flow gas is Ar gas (99.99%). The $CrCl_3 \cdot 6H_2O$ powders were put in the upstream zone of the furnace, where the temperature ranged from 700°C to 800°C, while the silicon (001) substrates were put in the downstream zone with the same temperature range. During the growth process, with oxygen environment, $CrSi_2$ nanowires may transform to be $CrSi_2$ (core)/ SiO_2 (shell) nanowires due to oxidation. To understand what factors influence the growth of chromium disilicide nanowires, we varied reaction time and temperatures of substrates and the metal source. Scanning electron microscopy (SEM), X-ray diffraction (XRD), and transmission electron microscopy (TEM) studies were conducted for morphology observation and structure identification of the nanowires. Additionally, physical properties, including magnetism (SQUID), photoluminescence (PL), and field emission (Keithley-237), were measured.

RESULTS AND DISCUSSION

In this work, we controlled different parameters to realize how they influence the nanowires' growth, morphology, and physical properties. With source and substrate at 700°C and the flow gas of 120 sccm, we obtained dense $CrSi_2$ nanowires with a length of approximately 20 μm as shown in Figure 1a by chemical vapor deposition. Interestingly, in Figure 1b, the nanowires grew from the particle with almost coherent growth direction and the morphology was rare. XRD analysis in Figure 1c shows (111), (003), and (112) major plane peaks, indicating that the nanowires have a C40 hexagonal structure. The TEM image of Figure 2a shows that the nanowires are 10 to 50 nm in diameter. In Figure 2b, the high-resolution transmission electron microscopy (HRTEM) image and the corresponding fast Fourier transform (FFT) pattern in the inset identifies the materials to be single-crystal CrSi$_2$ nanowires of a hexagonal structure with lattice constants, $a = 0.4428$ nm and $c = 0.6369$ nm (JCPDS card no. 35–0781); the growth direction is [001], and the interplanar spacing of plane (003) is 0.2098 nm. Additionally, we tried 750°C with hydrogen as reducing atmosphere and obtained Cr$_5$Si$_3$ nanowires of approximately 10 μm in length and of a different morphology as shown in Figure 1d. In Figure 1e, we found that the nanowires grew from nanoparticles again. XRD analysis in Figure 1f shows two phases, CrSi$_2$ and Cr$_5$Si$_3$; for further investigation on the atomic structures of the nanowires, we conducted TEM analysis as shown in Figure 2. From the TEM image of Figure 2c, the nanowire was of approximately 80 nm in diameter. The HRTEM image and the corresponding FFT pattern in the inset of Figure 2d confirm that the single-crystal Cr$_5$Si$_3$ nanowire has a BCT D8m structure with lattice constants, $a = 0.9165$ nm and $c = 0.4638$ nm (JCPDS card no. 51–1357); also, the nanowire is with [100] growth direction, and the interplanar spacing of plane (200) is 0.4571 nm.

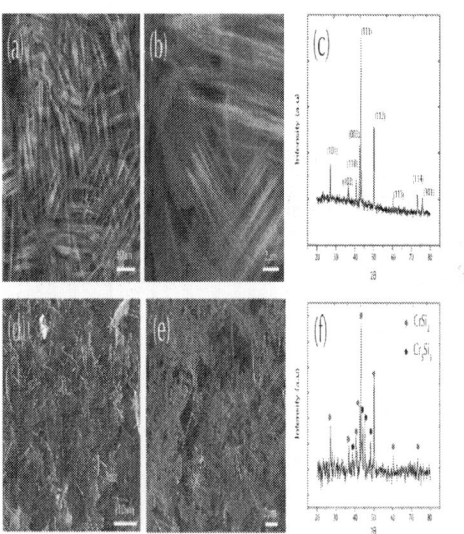

Figure 1: SEM images and XRD analysis of chromium silicide nanowires. (a) Low magnification, (b) high-resolution SEM images, and(c) XRD analysis of CrSi$_2$ nanowires grown at 700°C. (d) Low magnification, (e) high-resolution SEM images and (f) XRD analysis of Cr$_5$Si$_3$ nanowires grown at 750°C with H$_2$ atmosphere.

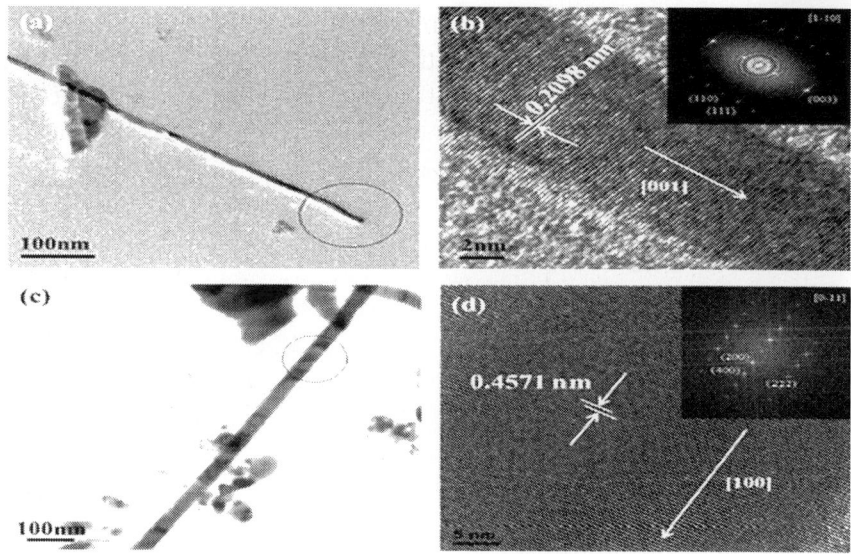

Figure 2: TEM analysis of chromium silicide nanowires. (a) Low magnification, (b) high-resolution TEM images of CrSi$_2$ nanowires grown at 700°C. The inset in (b) shows the corresponding fast Fourier transform (FFT) pattern with a zone axis of [1–10]. (c) Low magnification, (d) high-resolution TEM images of Cr$_5$Si$_3$ nanowires grown at 750°C. The inset in (d) shows the corresponding FFT pattern with a zone axis of [0–11].

The growth mechanism of the chromium silicide nanowires in this study is interesting. Figure 3 is the schematic illustration of the growth mechanism, showing the proposed growth steps of the CrSi$_2$ nanowires. When the system was heated below 700°C, CrCl$_3 \cdot$ 6H$_2$O transformed to CrCl$_3$ and H$_2$O:

$$CrCl_3.6H_2O_{(g)} \rightarrow CrCl_{3(g)} + 6H_2O$$

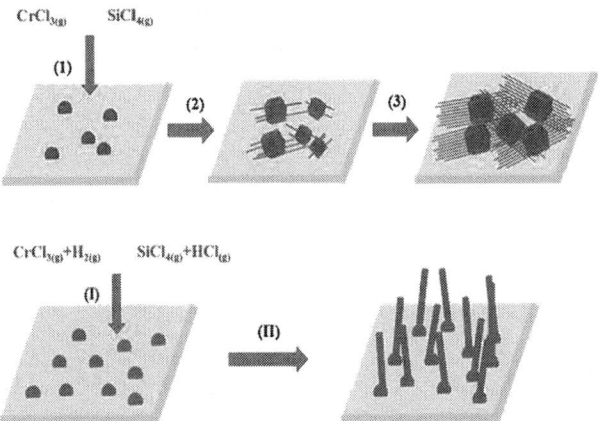

Figure 3: Schematic illustration of the growth mechanism. (1) $4CrCl_{3(g)}+11$ $Si_{(s)} \rightarrow 4CrSi_{2(s)} + 3SiCl_{4(g)}$; $4SiCl_{4(g)} + 2CrCl_{3(g)} \rightarrow 2CrSi_{2(l)} + 11Cl_{2(g)}$. (2) Growth of $CrSi_2$ particles and nanowires. (3) High-density $CrSi_2$ nanowires. (I)

$$10CrCl_{3(g)} + 12Si_{(s)} + 3H_{2(g)} \rightarrow 2Cr_5Si_{3(s)} + 6SiCl_{4(g)} + 6HCl_{(g)}.$$ (II) Growth of Cr_5Si_3 nanowires.

The $CrCl_3$ gas molecules then agglomerated on the silicon substrate. As the system temperature reached the reaction temperature, 700°C, $CrCl_3$ gas reacted with the silicon substrate to form $CrSi_2$ nanoparticles and $SiCl_4$ based on step (1) of Figure 3:

$$4CrCl_{3(g)} + 11Si_{(s)} \rightarrow 4CrSi_{2(s)} + 3SiCl_{4(g)} \quad T=700°C$$

The $SiCl_4$ product then reacted with $CrCl_{3(g)}$ to form $CrSi_2$, following step (2) of Figure 3:

$$4SiCl_{4(g)} + 2CrCl_{3(g)} \rightarrow 2CrSi_{2(l)} + 11Cl_{2(g)} \quad T=700°C$$

Notably, the $CrSi_2$ nanowires precipitated from polygonal particles, and the growth direction seems consistent as shown in Figure 1b. The nanowires and polygonal particles may have the same stacking plane, (003), based on our TEM analysis, and nanowires grew from voids and defects on the surface of any polygonal particles with <001> growth direction, following step (3) of Figure 3 as shown in a SEM image of Additional file 1: Figure S1. We conducted experiments with the heating times of 1.5, 4, and 12 h at 700°C, obtaining the corresponding results shown in Figure 4a, b, c, respectively. We found nanowires and particles at 1.5 h, more nanowires growing from particles at 4 h, and dense nanowires appearing with buried particles at 12 h, respectively. With a longer duration, more nanowires can overcome the activation energy, successfully nucleate, and grow to be nanowires, contributing to $CrSi_2$ nanowires of a high density. According to the observations, we proposed that the mechanism of the nanowire growth is a self-catalytic process.

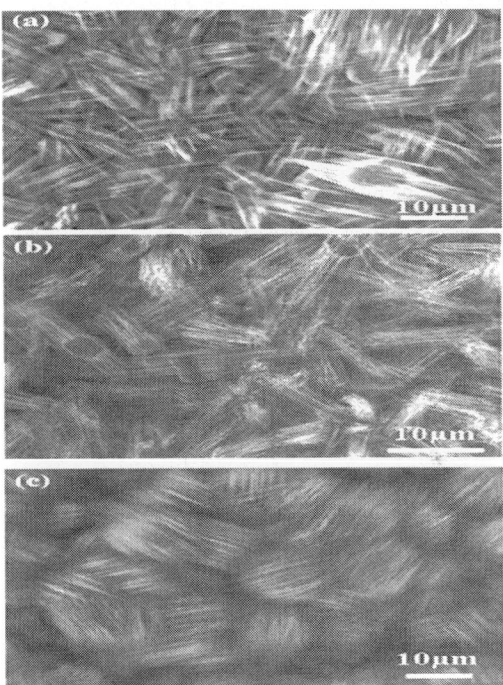

Figure 4: SEM images of $CrSi_2$ nanowires at different heating times of (a) 1.5, (b) 4, and (c) 12 h, respectively.

As the substrate temperature was at 750°C, $CrCl_3$ gas reacted with H_2 gas and the silicon substrate to form Cr_5Si_3 nanoparticles, HCl, and $SiCl_4$, following step (i) of Figure 3:

$$10CrCl_{3(g)}+12S_{i(s)}+3H_{2(g)}\rightarrow2Cr_5Si_{3(s)}+6SiCl_{4(g)}+6HCl_{(g)}, T=750°C$$

The $SiCl_4$ also reacted with $CrCl_3$ to form $CrSi_2$, which is the reason why the XRD analysis shows both $CrSi_2$ and Cr_5Si_3 phases.

Also, we investigated the influence of the carrier gas flow rate when synthesizing chromium silicide nanowires. We conducted experiments at the gas flow rate of 60, 120, and 240 sccm at 700°C, obtaining the corresponding results shown in Figure 5a, b, c, respectively. It can be found that chromium disilicide nanowires appeared without particles at 60 sccm and with few particles at 120 sccm and that the morphology gradually transformed from nanowires to films at 240 sccm.

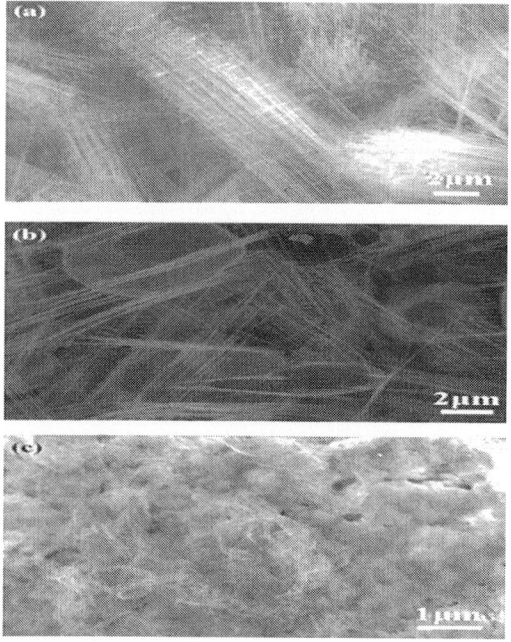

Figure 5: SEM images of $CrSi_2$ nanowires at different gas flow rates of (a) 60, (b) 120, and (c) 240 sccm, respectively.

The CVD synthesis system can be divided into three sub-systems, which are momentum control system, mass transfer control system, and surface reaction control system. At a lower gas flow rate, mass transfer control system would be the main reaction mechanism, with which gas adsorption and desorption occurred on the Si wafer and fabrication of chromium silicide nanowires was preferred. On the other hand, at a higher gas flow rate, surface reaction control system would be the main reaction mechanism, with which $CrCl_3$ reacted on the Si wafer surface by chemical vapor deposition; thus, chromium silicide films appeared.

In addition to understanding the growth behaviors of the chromium silicide nanowires, we explored their physical properties. Figure 6 is the field emission measurements for $CrSi_2$ NWs, showing the plot of the current density (J) as a function of the applied field (E) with the inset of the $\ln(J/E^2)$-1/E plot. The sample was measured in a vacuum chamber pump to approximately 10^{-6} Torr. According to the Fowler-Nordheim (F-N) plot and the Fowler-Nordheim equation:

$$=(A\beta^2 E^2/\varphi)\exp(-B\varphi^{3/2}/\beta E)$$

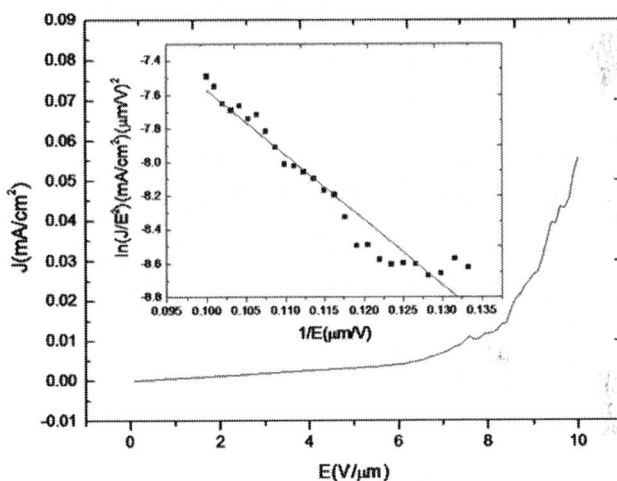

Figure 6: The field emission measurements of $CrSi_2$NWs; the inset shows the corresponding $\ln (J/E^2)^{-1}$/E plot.

where J is the current density, E is the applied electric field, φ is the work function, and A, B are constants, respectively. We put +1,000 V on the sample with a 100-μm spacing between the anode and cathode, and we defined the turn-on field could obtain a current density of 10 μA/cm^2 and the turn-on field we measured for $CrSi_2$ nanowires was 7.5 V/μm. The field enhancement factor β has been calculated to be 1,366 from the slope of $\ln(J/E^2) = \ln(A\beta^2/\varphi) - B\varphi^{3/2}/\beta E$ (for $CrSi_2$ $\varphi = 3.9$ eV[19]), demonstrating that $CrSi_2$ NWs are promising emitters. The outstanding field emission properties of $CrSi_2$ NWs are attributed to their metallic property and special one-dimensional geometry with a high aspect ratio as compared with those of many other materials.

On magnetization analysis for chromium disilicide nanowires coated with a silicon oxide layer of a few nanometers in thickness, we prepared samples of 2.5 mm × 2.5 mm with the applied magnetic field of ±3,000 Oe perpendicular to the substrates. Notably, Figure 7 shows that the $CrSi_2/SiO_x$ nanowires grown here were found to be ferromagnetic with the saturation magnetization of 8×10^{-7} emu, M_R, remanence, of 2×10^{-7} emu, and H_C, coercive force, of about 179 Oe, respectively, which is different from the antimagnetic behavior in $CrSi_2$ and SiO_x. The ferromagnetic characteristic results from the bonding formation between the Si sp hybrid orbitals and the Cr 3d orbitals at the $SiOx/CrSi_2$ interface, where the oxygen atoms play an important role, bonding with silicon atoms and making chromium atoms with unpaired electrons, which contributes to ferromagnetism at nanoscale [25].

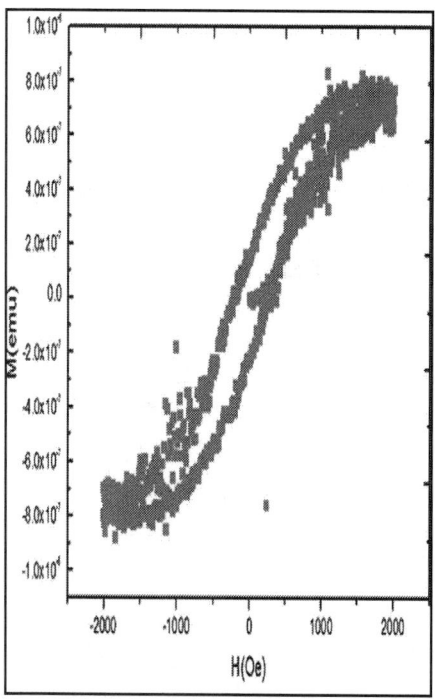

Figure 7: The magnetism measurements of $CrSi_2/SiO_x$ nanowires.

On photoluminescence analysis, Bhamu et al. studied the density of state (DOS) of CrSi$_2$ bulk, including 1.33 eV, 0.56 eV above Fermi state, and 2.23 eV under Fermi state [29]. Figure 8b shows our PL spectrum in the visible region for the CrSi$_2$ nanowires, where the wide peak was present (red line) and through Gaussian fitting; the other two peaks, 396 nm (green line) and 465 nm (blue line), were calculated. Theoretically, the electron-hole pair recombinations of 1.33 eV, 0.56 eV conduct state to −2.23 eV valance state were 348 and 430 nm for CrSi$_2$ bulk. In reality, the difference results from dimension, bulk, and nanowires; as the particle size reduces, wider bandgap light absorption band will move to shorter wavelengths, which is so-called blueshift [30]; however, there may be redshift as well; as the particle size decreases, the internal stress will increase, causing changes in the band structure [31] and the electron wave function overlap to increase the energy gap narrowing[32]; if the redshift factor is larger than the blueshift, then we will see redshift phenomenon, which is the case here.

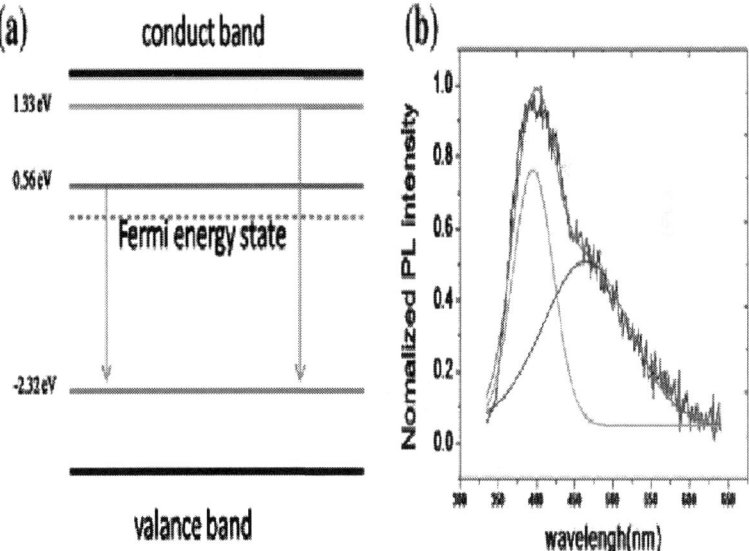

Figure 8: PL spectrum for the CrSi$_2$nanowires. (a) Energy states of CrSi2 bulk. (b) Photoluminescence measurements of CrSi2 NWs with Gaussian fitting.

CONCLUSIONS

In this study, using a CVD method, we have successfully synthesized chromium silicide nanowires of two phases with unique morphologies. Effects of some processing parameters, including the temperature, gas flow rate, and heating time, were investigated; for example, the growth of chromium disilicide nanowires were influenced by $CrSi_2$ vapor supersaturation, $CrSi_2$ vapor formation rate, and CVD control system. Also, the growth mechanism has been proposed. Field emission and photoluminescence measurements demonstrate that the $CrSi_2$ nanowires are potential field-emitting and photovoltaic materials with a low turn-on field. Additionally, the magnetic property measurements for the $CrSi_2/SiO_x$ nanowires, showing a ferromagnetic characteristic, demonstrate promising applications for magnetic storage and biological cell separation.

AUTHORS' CONTRIBUTIONS

HFH and KCL conceived the study and designed the research. HFH conducted the experiments. HFH, PCT, and KCL wrote the manuscript. All authors read and approved the final manuscript.

ACKNOWLEDGEMENTS

KCL acknowledges the support from the National Science Council through grants 100-2628-E-006-025-MY2 and 102-2221-E-006-077-MY3.

REFERENCES

1.   Lu CM, Hsu HF, Lu KC: Growth of single-crystalline cobalt silicide nanowires and their field emission property.*Nanoscale Res Lett* 2013, 8:308.

2.   Chiu WL, Chiu CH, Chen JY, Huang CW, Huang YT, Lu KC, *et al.*: Single-crystalline **δ**-Ni2Si nanowires with excellent physical properties.*Nanoscale Res Lett* 2013, 8:290.

3. Lu KC, Wu WW, Ouyang H, Lin YC, Huang Y, Wang CW, *et al.*: The influence of surface oxide on the growth of metal/semiconductor nanowires. *Nano Lett* 2011, 7:2753-8.

4. Wu WW, Lu KC, Chen KN, Yeh PH, Wang CE, Lin YC, *et al.*: Controlled large strain of Ni silicide/Si/Ni silicide nanowire heterostructures and their electron transport properties. *Appl Phys Lett* 2011, 97:203110.

5. Wu WW, Lu KC, Wang CW, Hsieh HY, Chen SY, Chou YC, *et al.*: Growth of multiple metal/semiconductor nanoheterostructures through point and line contact reactions. *Nano Lett* 2010, 10:3984-9.

6. Chou YC, Lu KC, Tu KN: Nucleation and growth of epitaxial silicide in silicon nanowires. *Mat Sci Eng R* 2010, 70:112-25.

7. Lu KC, Wu WW, Wu HW, Tanner CM, Chang JP, Chen LJ, *et al.*: In situ control of atomic-scale Si layer with huge strain in the nanoheterostructure NiSi/Si/NiSi through point contact reaction. *Nano Lett* 2007, 8:2389-94.

8. Lu KC, Tu KN, Wu WW, Chen LJ, Yoo BY, Myung NV: Point contact reactions between Ni and Si nanowires and reactive epitaxial growth of axial nano-NiSi/Si. *Appl Phys Lett* 2007, 90:253111.

9. Liang YH, Yu SY, Hsin CL, Huang CW, Wu WW: Growth of single-crystalline cobalt silicide nanowires with excellent physical properties. *J Appl Phys* 2011, 110:074302.

10. Chen LJ: An integral part of microelectronics. *JOM* 2005, 57:24-30.

11. Derrien J, Chevrier J, Lethanh V, Mahan JE: Semiconducting silicide-silicon heterostructures: growth, properties and applications. *Appl Surf Sci* 1992, 382:56-8.

12. Ng WL, Lourenco MA, Gwilliam RM, Ledain S, Shao G, Homewood KP: An efficient room-temperature silicon-based light-emitting diode. *Nature* 2001, 410:192-4.

13. Leong D, Harry M, Reeson KJ, Homewood KP: A silicon/iron-disilicide light-emitting diode operating at a wavelength of 1.5 mum. *Nature* 1997, 387:686-8.

14. Bost MC, Mahan JE: An investigation of the optical constants and band gap of chromium disilicide. *J Appl Phys* 1988, 63:839-44.

15. Shinoda D, Asanabe S, Sasaki Y: Semiconducting properties of chromium disilicide.*J Phys Soc Jpn* 1964, 19:269-72.

16. Bellani V, Guizzetti G, Marabelli F, Piaggi A, Borghesi A, Nava F, et al.: Theory and experiment on the optical properties of $CrSi_2$. *Phys Rev B* 1992, 46:9380-9.

17. Mattheiss LF: Electronic structure of $CrSi_2$ and related refractory disilicides.*Phys Rev B* 1991, 43:12549-55.

18. Anderson WA, Delahoy AE, Milano RA: 8 percent efficient layered Schottky-barrier solar cell.*J Appl Phys* 1974, 45:3913-5.

19. Bewlay BP, Lipsitt HA, Jackson MR, Chang KM: Processing microstructures and properties of Cr-Cr sub 3 Si, Nb-Nb sub 3 Si, and V-V sub 3 Si eutectics.*Mater Manuf Processes* 1994, 9:89-109.

20. Nishida I: The crystal growth and thermoelectric properties of chromium disilicide.*J Mater Sci* 1972, 7:1119-24.

21. Rowe DM: *CRC handbook of thermoelectrics*. CRC Press, Boca Raton, FL; 1995.

22. Chung IJ, Hariz A: Surface application of metal silicides for improved electrical properties of field-emitter arrays.*Smart Mater Struct* 1997, 6:633-9.

23. Bonard JM, Salvetat JP, Stockli T, Forro L, Chatelain A: Field emission from carbon nanotubes: perspectives for applications and clues to the emission mechanism.*Appl Phys A* 1999, 69:245-54.

24. Minami T, Miyata T, Yamamoto T: Transparent conducting zinc-co-doped ITO films prepared by magnetron sputtering.*Surf Coat Technol* 1998, 108:583-7.

25. Hou TC, Han YH, Lo SC, Lee CT, Ouyang H, Chen LJ: Room-temperature ferromagnetism in $CrSi_2$(core)/SiO_2(shell) semiconducting nanocables.*Appl Phys Lett* 2011, 98:193104.

26. Zhang Y, Wu Q, Qian W, Liu N, Qin X, Yu L, et al.: Morphology-controlled growth of chromium silicide nanostructures and their field emission properties.*CrystEngComm* 2012, 14:1659-64.

27. Seo K, Varadwaj KSK, Cha D, In J, Kim J, Park J, et al.: Synthesis and electrical properties of single crystalline $CrSi_2$ nanowires.*J Phys Chem C* 2007, 111:9072-6.

28.   Lee CT, Li TY, Chiou SH, Lo SC, Han YH, Ouyang H: First-principles analyses of unusual ferromagnetism observed in $CrSi_2$(core)/ SiO_2(shell) nanocables. *J Appl Phys* 2013, 113:17E140.

29.   Bhamu KC, Sahariya J, Ahuja BL: Electronic structure of ceramic CrSi2 and WSi2: Compton spectroscopy and ab-initio calculations. *J Phys Chem Solids* 2013, 74:765-71.

30.   Wang Y, Herron N: Nanometer-sized semiconductor clusters: materials synthesis, quantum size effects, and photophysical properties. *J Phys Chem* 1991, 95:525-32.

31.   Fu H, Zunger A: Electronic structure, surface effects, and the redshifted emission InP quantum dots. *Phys Rev B* 1997, 56:1496-508.

32.   Smith CA, Lee HWH, Leppert VJ, Risbud SH: Ultraviolet-blue emission and electron-hole states in ZnSe quantum dots. *Appl Phys Lett* 1999, 75:1688-90.

Chapter 3

Chemical and Magnetic Functionalization of Graphene Oxide as a Route to Enhance its Biocompatibility

Karolina Urbas[1], Malgorzata Aleksandrzak[1], Magdalena Jedrzejczak[2], Malgorzata Jedrzejczak[2], Rafal Rakoczy[3], Xuecheng Chen[1], and Ewa Mijowska[1]

[1]Institute of Chemical and Environment Engineering, West Pomeranian University of Technology, Szczecin, Piastow Avenue 45, Szczecin 70-311, Poland

[2]Laboratory of Cytogenetics, West Pomeranian University of Technology, Szczecin, Judyma 6, Szczecin 71-466, Poland

[3]Institute of Chemical Engineering and Environmental Protection Process, West Pomeranian University of Technology, Szczecin, Piastow Avenue 42, Szczecin 71-065, Poland

ABSTRACT

The novel approach for deposition of iron oxide nanoparticles with narrow size distribution supported on different sized graphene oxide was reported. Two different samples with different size distributions of graphene oxide (0.5 to 7 µm and 1 to 3 µm) were selectively prepared, and the influence of the flake size distribution on the mitochondrial activity of L929 with WST1 assay in vitro study was also evaluated. Little reduction of mitochondrial activity of the $GO-Fe_3O_4$ samples with broader size distribution (0.5 to 7 µm) was observed. The pristine GO samples (0.5 to 7 µm) in the highest concentrations reduced the mitochondrial activity significantly. For $GO-Fe_3O_4$ samples with narrower size distribution, the best biocompatibility was noticed at concentration 12.5 µg/mL. The highest reduction of cell viability was noted at a dose 100 µg/mL for GO (1 to 3 µm). It is worth noting that the chemical functionalization of GO and Fe_3O_4 is a way to enhance the biocompatibility and makes the system independent of the size distribution of graphene oxide.

BACKGROUND

In recent years, graphene, well-defined 2D honeycomb-like network of carbon atoms, has attracted growing interest owing to its unprecedented combination of unique electrical, thermal, optical, and mechanical properties [1-6].

Graphene derivative, graphene oxide chemically exfoliated from oxidized graphite, is considered as a promising material for biological applications due to its surface functionalizability, amphiphilicity, and excellent aqueous processability. These extraordinary properties are mainly derived from its chemical structures composed of sp^3 carbon domains surrounding sp^2 carbon domain and a wide range of functional groups such as epoxy, hydroxyl, and carboxyl groups [7-10]. The chemical structure of graphene oxide and large specific surface area enable various chemical modification or functionalization and make graphene oxide an excellent platform for loading magnetic nanoparticles [11].

Magnetic nanoparticles possessing tailored surface properties and appropriate physicochemistry have been widely investigated for various applications such as hyperthermia, magnetic resonance imaging (MRI), tissue repair, drug delivery, biosensing, and bioanalysis [12-23]. In particular, the magnetite, Fe_3O_4, has attracted significant attention in the field of biotechnology and medicine because of its strong magnetic properties and low toxicity [24-26]. The properties of nanocrystals strongly depend on the dimension of the nanocrystals; therefore, the control of monodispersed size of nanocrystals plays an important role. Magnetic nanoparticles for the use in biomedical applications are desired to exhibit superparamagnetic properties. The superparamagnetic nature implies that the particles will not be attracted to each other, and so the risk of agglomeration in a medical setting is minimized. Magnetite is traditionally ferromagnetic in nature. However, as the size decreases to 30 nm or smaller, it loses their permanent magnetism and becomes superparamagnetic [27]. Safety concerns could ultimately prevent the adoption of magnetic nanoparticles in medicine. In vitro and in vivo toxicity results often contradict each other hence are an area that needs more research.

Recently, graphene-based materials were extensively investigated for application in biosensing[28-32], imaging [33,34], and drug delivery [35-38] as vehicles for drugs and as high-performance electrode material for capacitive deionization [39,40]. Cong et al. report on fabrication of reduced graphene oxide decorated with Fe_3O_4 nanoparticles through a high-temperature decomposition method [41]. This system could be used as magnetic resonance contrast agent. Shen et al. demonstrated one-step synthesis of GO-Fe_3O_4 nanoparticle hybrid [42]. He and Gao presented scalable, green, efficient, controllable method of preparation of superparamagnetic, processable, and conductive graphene nanosheets coated with magnetite nanoparticles [43]. He et al. showed attachment of magnetite nanoparticles to GO surface with covalent bonding [44]. Yang et al. described GO-Fe_3O_4 nanoparticle hybrid supporting doxorubicin hydrochloride (anticancer drug) [35]. This system could be easily removed from water by an external magnetic field. Zheng and Li reported on fabrication of a magnetite nanoparticle-decorated graphene oxide (Fe_3O_4-GO) and reduced graphene oxide (Fe_3O_4-rGO) loaded with β-lapachone (anticancer drug), in vitro anticancer efficacy and cytotoxicity of obtained materials[45]. Bai et al. presented results of study on the inductive heating property of graphene oxide sheets decorated with magnetite nanoparticles in AC magnetic field [46]. The potential of the obtained nanocomposite was evaluated for localized hyperthermia treatment of cancer cells.

Herein, we present new facile approach for production of the monodispersed Fe_3O_4 nanoparticles and magnetic attachment of magnetite nanoparticles to graphene oxide sheets with different flake size distributions. The mean size of the obtained magnetite nanoparticles is about 8 nm. Additionally, we performed cytocompatibility study on the influence of these molecular hybrids on the mitochondrial activity of L929 cell line with WST1 assay in respect to the GO and pristine iron oxide nanoparticles. The cellular response was verified with different concentration (0.0, 3.125, 6.25, 12.5, 25.0, 50.0, 100.0 µg/mL) of the nanomaterials.

METHODS

Preparation of Graphene Oxide-Fe₃O₄ Nanoparticle Hybrid

Synthesis of Graphene Oxide

Two types of samples of graphene oxide were synthesized by oxidation of graphite with various size of flakes (with narrow and broad size distribution) using the modified Hummer's method fully described elsewhere [47]. To a mixture of 6 g $KMnO_4$ and 1 g graphite, 120 mL of concentrated sulfuric acid and 15 mL of orthophosphoric acid were poured. It was heated to 50°C and stirred for 24 h. The resulting mixture was added to ice (150 mL) with 1 mL of H_2O_2 (30%) and centrifuged. The separated solid product was washed two times with water and 30% HCl and ethanol and left for vacuum drying for 12 h at 70°C. The sample with broad size distribution is named B-GO, and the sample with narrow size distribution is named N-GO.

Synthesis of Magnetite Nanoparticles

Magnetic Fe_3O_4 nanoparticles were synthesized by co-precipitation of Fe^{2+} and Fe^{3+} aqueous salt solutions using $NH_3 \cdot H_2O$ as the precipitating agent in order to adjust the pH value. It should be noticed that the size, shape, and composition of nanoparticles may be controlled by means of the type of salts used, Fe^{2+} and Fe^{3+} ratio, pH, and ionic strength of the media. To synthesize Fe_3O_4 nanoparticles, the solutions of 3.9 g Mohr's salt $(NH_4)_2Fe(SO_4)_2 \cdot 6H_2O$ in 100 mL H_2O (0.1 M) and 4.8 g $NH_4Fe(SO_4)_2 \cdot 12H_2O$ in 100 mL H_2O (0.1 M) were prepared and mixed with a molar ratio of 1:2. Ammonia aqueous solution was dropped into the mixture slowly until the pH value of the solution reached 9. The complete precipitation of Fe_3O_4 was expected between pH 9 and 14. The overall reaction may be written as follows:

$$(NH_4)_2Fe(SO_4)_2 \cdot 6H_2O \; + \; 2\,NH_4Fe(SO_4)_2 \cdot 12H_2O$$
$$+ \; 8\,NH_4OH \rightarrow$$
$$\rightarrow Fe_3O_4 \cdot 4H_2O \; + \; 6\,(NH_4)_2SO_4 + \; 14\,H_2O$$

The magnetite synthesis route was carried out under the action of a rotating magnetic field (RMF). A liquid-filled glass container was placed inside the three-phase stator of an induction squirrel-cage motor which generated the RMF. This kind of the magnetic field might be used to augment the process intensity instead of a mechanical mixing. One of the advantages of the RMF is the possibility to apply it to generation and control of the hydrodynamic states for the magnetic particle mixing systems. In the experimental procedure, the frequency of the RMF was equal to 50 Hz. The intensity of the magnetic field could be 25 mT. The more detailed information about the experimental setup and the measurements of the magnetic field for the tested apparatus may be found here [48]. Finally, the precipitate was collected by filtration and washed three times with deionized water and then dried.

Synthesis of Graphene Oxide-Fe$_3$O$_4$ Nanoparticle Hybrid

The synthesis process of GO-Fe$_3$O$_4$ nanocomposite is schematically illustrated in Figure 1.Twenty milligrams of each of graphene oxide sample (B-GO and N-GO) was exfoliated in 60 mL H$_2$O by ultrasonication to produce a homogeneous graphene oxide water-based suspension. Then, the carboxylic groups on the graphene oxide surface were activated with 8 mg of N-hydroxysuccinimide (NHS) and 10 mg of 1-(3-dimethylaminopropyl)-3-ethylcarbodiimide (EDC). The surface of iron oxide (20 mg) was modified with oleic acid (3 mL) and then stirred and sonicated for 1 h. The mixture of modified iron oxide and graphene oxide was stirred for 48 h. In the next step, the mixture was filtered by a polycarbonate membrane and washed several times with water and ethanol. Finally, the obtained product was dried for 12 h at 100°C. The sample with broad size distribution of graphene oxide flakes is named B-GO-Fe$_3$O$_4$, and the sample with narrow size distribution is named N-GO-Fe$_3$O$_4$.

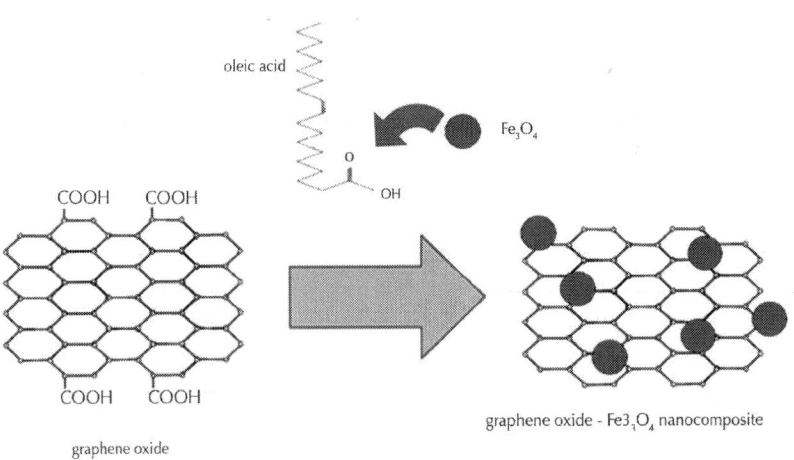

Figure 1: Schematic of the synthesis of GO-Fe$_3$O$_4$nanoparticle hybrid.

Characterization

High-resolution transmission electron microscopy (HRTEM) (FEI Tecnai F30, Frequency Electronics Inc., Mitchel Field, NY, USA) was employed to examine the morphology of the samples and the size and distribution of the magnetite nanoparticles. X-ray diffraction technique (X-ray diffractometer Philips X'Pert PRO, PANalytical B.V., Almelo, The Netherlands, $K\alpha_1 = 1.54056$ Å) was used to investigate the structure of the samples and to estimate the average size of magnetite nanoparticles. In order to study the thickness of obtained graphene oxide flakes and nanocomposite, atomic force microscopy (Nanoscope V Multimode 8, Bruker AXS, Mannheim, Germany) was employed. IR absorption spectra were collected on the Nicolet 6700 FTIR spectrometer (Thermo Nicolet Corp., Madison, WI, USA). In order to investigate the thermal behavior of the samples, thermogravimetric analysis was performed on the SDT Q600 simultaneous TGA/DSC (TA Instruments Inc., Milford, MA, USA) under an air flow of 100 mL/min at heating rate of 5°C/min. Raman spectra were acquired on the inVia Raman Microscope (Renishaw PLC, New Mills Wotton-under-Edge, Gloucestershire, UK) at an excitation wavelength of 785 nm.

Cell Culture

The cell line of mouse fibroblasts (L929) were seeded on the 96-well plates at the density of 7.4×10^3 per well. Cells were maintained using DMEM cell culture medium (Gibco Corp., Grand Island, NY, USA) supplemented with 10% heat-inactivated fetal bovine serum (Gibco Corp., Grand Island, NY, USA), 0.4% streptomycin/penicillin (Sigma-Aldrich Corp., St. Louis, MO, USA), and 2 mM L-glutamine (Sigma-Aldrich Corp., St. Louis, MO, USA) at 37°C, 5% CO_2, and 95% humidity. The 200 μL/well final volume of culture medium was used in experiment.

The Cytocompatibility Study

The cytocompatibility of nanomaterials was tested using WST-1 Cell Proliferation Assay (Roche Applied Science, Penzberg, Germany) [49]. The test principle is based on the transformation of WST-1 salt [2-(4-iodophenyl)-3-(4-nitrophenyl)-5-(2,4-disulfophenyl)-2H-tetrazolium] into water-soluble colored formazan by mitochondrial dehydrogenases [50] that are active in rapidly dividing cells [51]. The generation of the dark yellow colored formazan is directly correlated to the number of the metabolically active cells; therefore, the cell number can be quantified by the photometric detection of the formazan. There are several similar proliferation assays using other tetrazolium salts, such as MTT [3-(4,5-dimethylthiazol-2-yl)-2,5-diphenyltetrazolium bromide], XTT [2,3-bis(2-methoxy-4-nitro-5-sulfophenyl)-2H-tetrazolium-5-carboxanilide], and MTS [3-(4,5-dimethylthiazol-2-yl)-5-(3-carboxymethoxyphenyl)-2-(4-sulfophenyl)-2H-tetrazolium] available on the market. The main advantage of WST-1 test over those mentioned above is the solubility of reduced WST-1 salt. It also requires no washing, harvesting, or solubilization of cells. To perform the assay, L929 cells were plated in the 96-well plates for 24 h. After incubation period from cells seeding, N-GO, B-GO, N-GO-Fe_3O_4, B-GO-Fe_3O_4, and Fe_3O_4 were introduced separately to cells with different final concentrations (0.0, 3.125, 6.25, 12.5, 25.0, 50.0, 100.0 µg/mL) in culture medium. Cells were incubated with nanomaterial for 24 h. Cells maintained in prepared medium without adding tested samples were taken as a control. To each well, 20 µL of WST-1 solution was added and incubated for additional 30 min at 37°C. After incubation, the absorbance at 450 nm, according to manufacturer's instructions, was recorded on the Sunrise Absorbance Reader (Tecan Group Ltd., Männedorf, Switzerland). All of the experiments were conducted in triplicate.

Statistical Analysis

All experiments were repeated at least three times. The results are given in the form: mean values ± standard deviation (SD). All results were compared using Student's t-test. Differences are considered significant at a level of $p < 0.05$.

RESULTS AND DISCUSSION

Transmission electron microscopy (TEM) was used for characterization of starting materials and final products. Representative TEM images of

Fe_3O_4, B-GO-Fe_3O_4, and N-GO-Fe_3O_4 nanoparticle hybrid are shown in Figure 2.Images of magnetite indicate a spherical shape of magnetite nanoparticles. The histogram presenting diameter distribution of the Fe_3O_4 nanoparticles is placed as an inset of Figure 2A. The diameter of the nanoparticles is in the range of 5 to 14 nm with strong peak at 8 nm. Uniform coverage of graphene oxide with magnetite nanoparticles can be observed in the images of the obtained nanocomposites (Figure 2B,C).

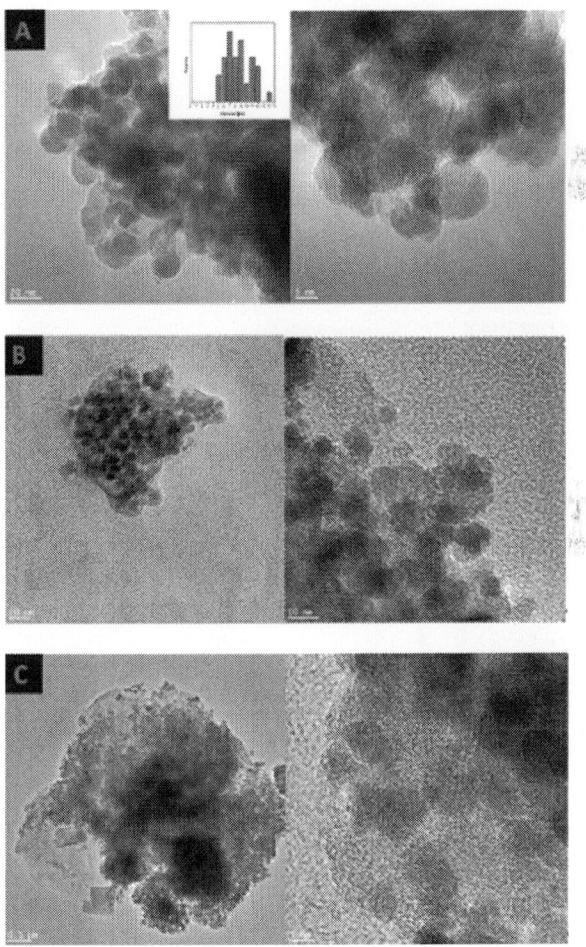

Figure 2: TEM images of Fe_3O_4 with diameter size distribution (A), B-GO-Fe_3O_4(B), and N-GO-Fe_3O_4(C).

To examine the morphology of the samples and the thickness of graphene oxide flakes, atomic force microscopy was employed. Atomic force microscopy (AFM) results of B-GO, N-GO, and B-GO-Fe_3O_4 and N-GO-Fe_3O_4 nanoparticle hybrid are presented in Figure 3. They indicate the change in the thickness of the flakes before and after functionalization - from 2 to 18 to 30 nm in both kinds of hybrids due to the deposition of the nanoparticles on graphene oxide flakes. AFM characterization is a useful tool for estimating the size distribution of the GO flakes. Figure 4A presents broad size distribution of graphene oxide flakes (from 0.5 to 7 μm) whereas Figure 4B presents narrower size distribution (from 1 to 3 μm).

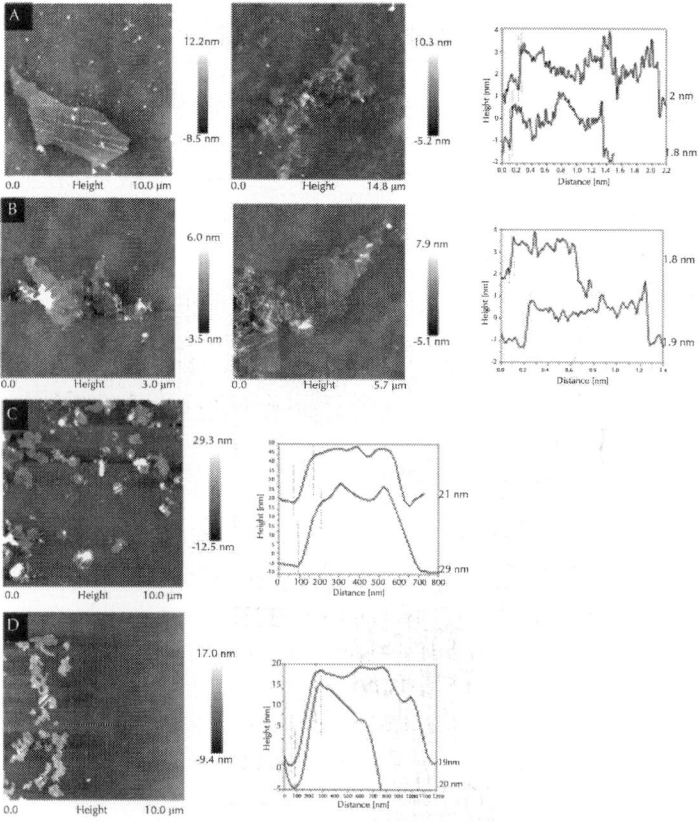

Figure 3: AFM images and height profiles of B-GO (A), N-GO (B), B-GO-Fe_3O_4(C), and N-GO-Fe_3O_4(D).

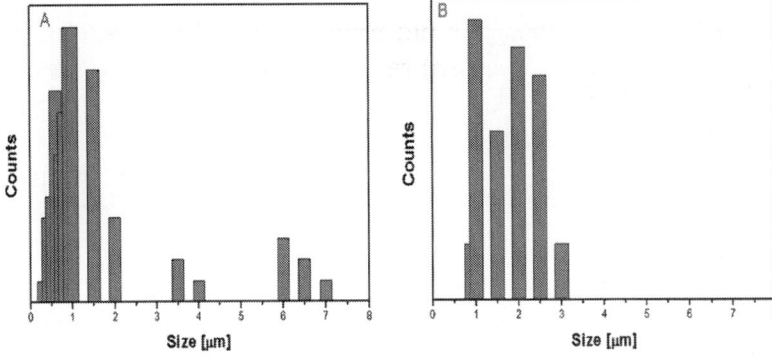

Figure 4: The size distributions of graphene oxide flakes in B-GO (A) and N-GO (B).

In order to follow the efficiency of the covalent functionalization of graphene oxide and iron oxide nanoparticle, the Fourier transform infrared (FTIR) spectroscopy was employed. Figure 5 presents FTIR spectra of GO, Fe_3O_4, B-GO-Fe_3O_4, and N-GO-Fe_3O_4. Figure 5A shows spectrum of GO with following peaks: 1,078 cm^{-1} corresponding to C-O stretching vibration mode related to the presence of the alkoxy group, 1,180 cm^{-1} attributed to C-O from the epoxy group, 1,475 cm^{-1} from C-OH carboxyl group, 1,625 cm^{-1} associated with the presence of C=C bond, and 1,742 cm^{-1} assigned to C=O stretching vibration mode in carboxyl group. Peak at 2,952 cm^{-1} originates from C-H bond. This spectrum confirms the presence of alkoxy, epoxy, and carboxyl group characteristic to graphene oxide. Figure 5B depicts FTIR absorption spectrum of Fe_3O_4 dominated by peaks at 570 and 1,059 cm^{-1} assigned to the presence of Fe-O bonds typical to magnetite. In the spectra of B-GO-Fe_3O_4 and N-GO-Fe_3O_4 (Figure 5C,D, respectively) peak from Fe-O bond has shifted from 570 to 809 cm^{-1} after functionalization. New peaks at 698 and 1,259 cm^{-1} occurred. The first one is attributed to N-H bond, and the second one is corresponded to C-N bond. Therefore, the presented spectra provide the proof for successful functionalization of GO with Fe_3O_4 via linkage of the oleic acid surrounding the iron oxide and NHS bonded to GO. The peak originating from the functional groups of graphene oxide are also present in the spectra of B-GO-Fe_3O_4 and N-GO-Fe_3O_4 which means that the starting material (GO) did not undergo the reduction during the functionalization.

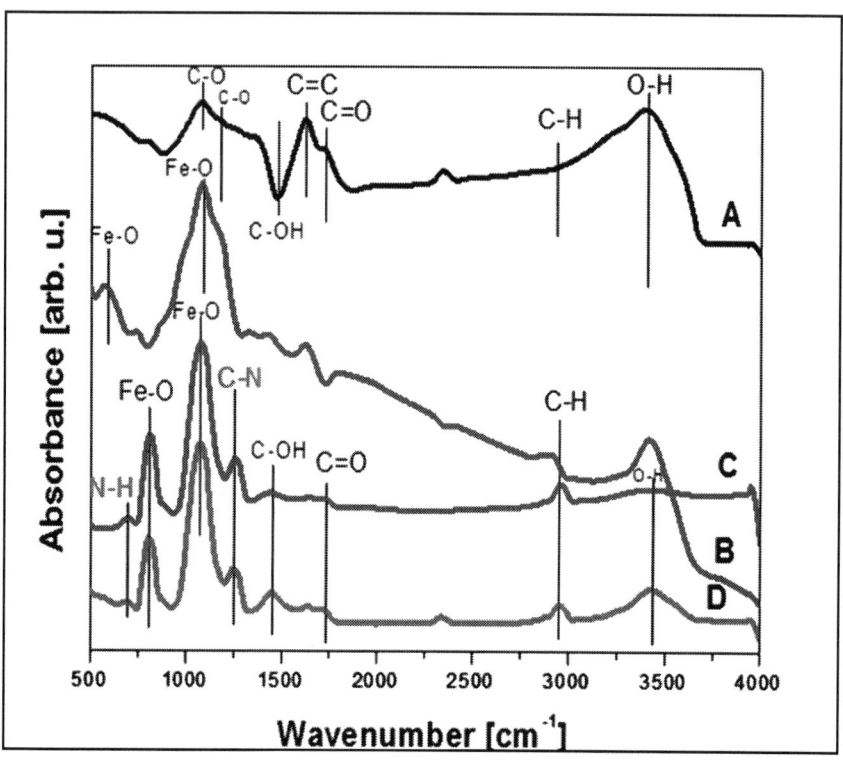

Figure 5: FTIR spectra of GO (A), Fe_3O_4(B), B-GO-Fe_3O_4(C), and N-GO-Fe_3O_4(D).

X-ray diffraction (XRD) patterns of graphite, graphene oxide, B-GO-Fe_3O_4, and N-GO-Fe_3O_4 hybrid are shown in Figure 6. XRD spectrum of graphite is dominated by intense and narrow peak at $2\theta=26.48°$ corresponding to reflection in (002) planes of well-ordered graphene layers. The lack of this reflection in the diffractogram of graphene oxide and the difference in the layer distances of the starting material and the final product confirm the completion of the oxidation reaction. Positions of the peaks and their relative intensities shown in XRD pattern of Fe_3O_4, B-GO-Fe_3O_4, and N-GO-Fe_3O_4 nanoparticle hybrid (Figure 6C,D,E) are consistent with the standard XRD data for magnetite (ICSD 65339). The average size of magnetite nanoparticles, calculated from the Scherrer's equation [52], is about 8 nm, which is consistent with the from TEM observations.

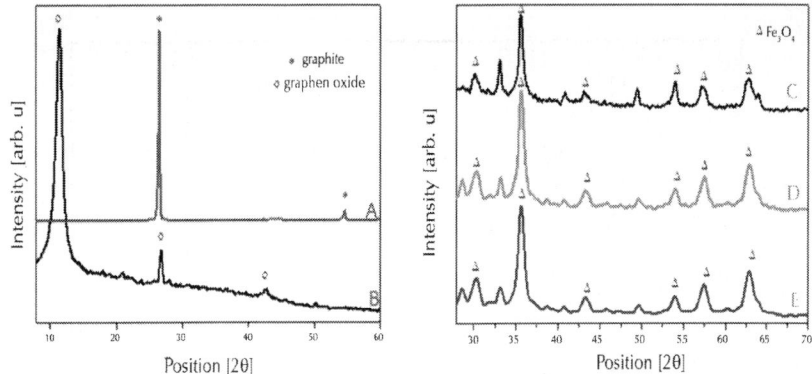

Figure 6: XRD patterns of graphite (A), GO (B), Fe_3O_4(C), B-GO-Fe_3O_4(D), and N-GO-Fe_3O_4(E).

Thermal gravimetric analysis is a useful tool to determine the composition of the samples. Figure 7 presents thermogravimetric (TG) curves of graphite, GO and B-GO-Fe_3O_4 and N-GO-Fe_3O_4 heated in the air Weight loss observed at 100°C is associated with the decomposition of physically adsorbed water. In graphene oxide (Figure 7B), the next weight loss occurred between 150°C and 300°C and it can be assigned to the decomposition and evaporization of oxygen-containing functional groups. After reaching 500°C (in graphite at 700°C - Figure 7A), the carbon skeleton underwent bulk pyrolysis. TG curves of nanocomposites (Figure 7C,D) indicate that the amount of loaded Fe_3O_4 is approximately 43 wt.% for B-GO-Fe_3O_4 and 40% for N-GO-Fe_3O_4. Both curves show the weight loss between 150°C and 300°C which means that graphene oxide did not underwent the reduction during the deposition of magnetite nanoparticles.

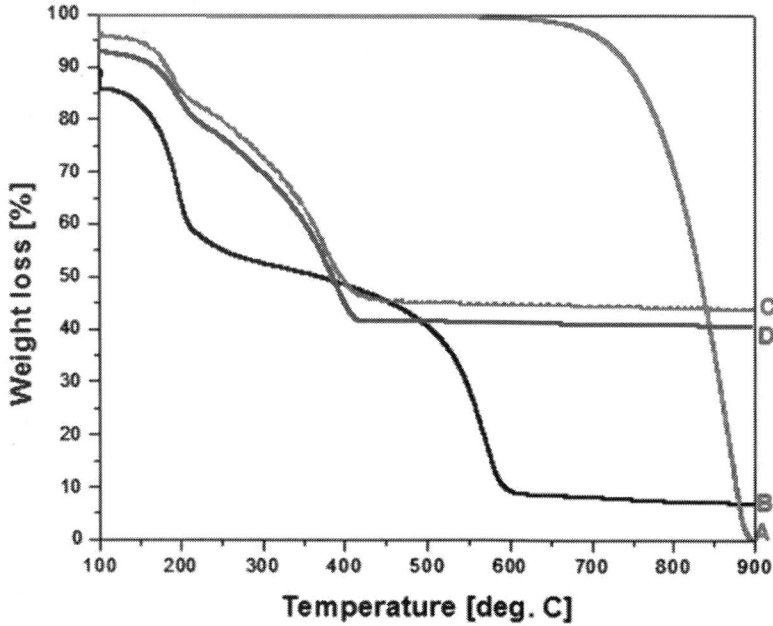

Figure 7: TG curves of graphite (A), GO (B), B-GO-Fe$_3$O$_4$(C), and N-GO-Fe$_3$O$_4$(D).

Raman spectroscopy is a powerful nondestructive tool to characterize carbonaceous materials, particularly for distinguishing ordered and disordered crystal structures of carbon. The significant structural changes occurring during the functionalization of GO are also reflected in their Raman spectra. Figure 8 presents Raman spectra of graphene oxide, B-GO-Fe$_3$O$_4$ and N-GO-Fe$_3$O$_4$. All spectra exhibit peaks at 1,318 and 1,602 cm^{-1} corresponded to D and G bands, respectively. The D band is associated with a certain fraction of sp^3 carbon atoms obtained due to an amorphization of graphite during the oxidation process, whereas the G mode originates from the in-plane vibration of sp^2 carbon atoms [53,54]. The ratio of ID/IG increases after functionalization of graphene oxide with magnetite nanoparticles in the case of both hybrid systems indicating successful functionalization of the starting material. Due to the presence of Fe$_3$O$_4$ nanoparticles in nanocomposites in the spectra of B-GO-Fe$_3$O$_4$ (Figure 8B) and N-GO-Fe$_3$O$_4$ (Figure 8C) additional peaks in the range of 200 to 300 cm^{-1} can be observed.

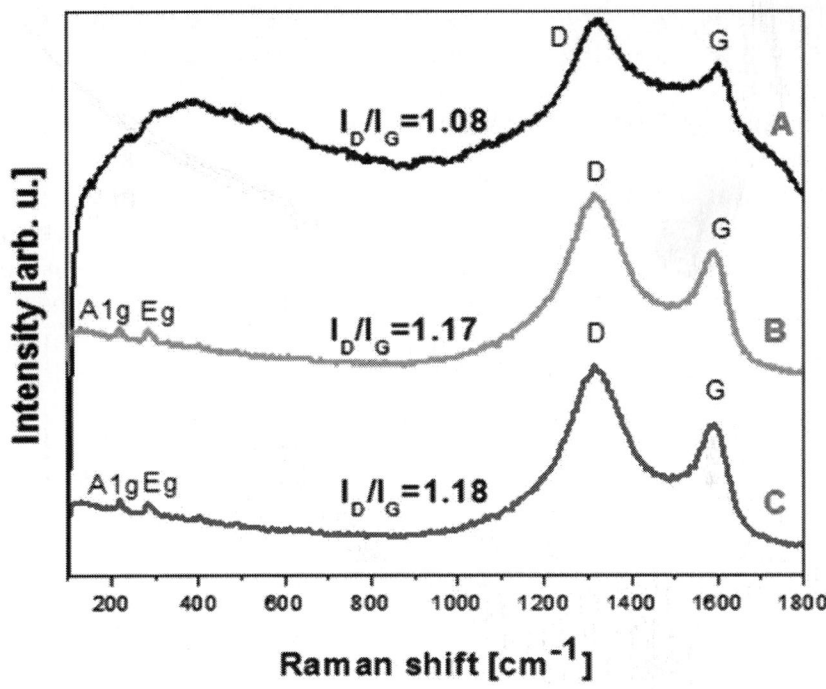

Figure 8: Raman spectra of GO (A), B-GO-Fe$_3$O$_4$(B), and N-GO-Fe$_3$O$_4$(C).

The good biocompatibility and safety of nanomaterials is fundamental for its medical application. The first step of this kind of investigations is in vitro analyses, e.g., mitochondrial activity. Here, five different samples have been examined for their cytocompatibility with L929 mouse fibroblasts: two samples of the nanocomposites of graphene oxide-Fe$_3$O$_4$ nanoparticles with different size distributions of GO flakes: B-GO-Fe$_3$O$_4$ and N-GO-Fe$_3$O$_4$, two samples of reference graphene oxide (B-GO and N-GO) and Fe$_3$O$_4$ nanoparticles. The presented biocompatibility study shows the differences in mitochondrial activity of L929 cells that depend on the type of nanomaterial and concentration (0.0, 3.125, 6.25, 12.5, 25.0, 50.0, 100.0 µg/mL). As shown in Figure 9(mitochondrial activity of the samples with broader size distribution), no significant cytotoxicity of the GO-Fe$_3$O$_4$ sample was detected for B-GO-Fe$_3$O$_4$. At the concentration between 3.125 and 100 µg/mL, minimal reduction of the cell mitochondrial activity was observed. Little reduction of cell viability was noted at a dose of 3.125 µg/mL in B-GO-Fe$_3$O$_4$ sample. In the sample with 25 µg/mL concentration, the mitochondrial activity for L929 cells was the highest. Interestingly, mitochondrial activity of the cells when interacting with the pristinegraphene oxide in the concentration between 50 and 100 µg/mL is significantly reduced. At a dose of 50 µg/mL, mitochondrial activity was reduced by approximately 40% for B-GO material. For 100 and 3.125 µg/mL concentration, cell viability was reduced to approximately 80%. Lower cytotoxicity was observed at a dose of 6.25 µg/mL in cell cultures treated with this graphene oxide. It means that the chemical functionalization of GO and deposition of Fe$_3$O$_4$ enhance the biocompatibility of the system.

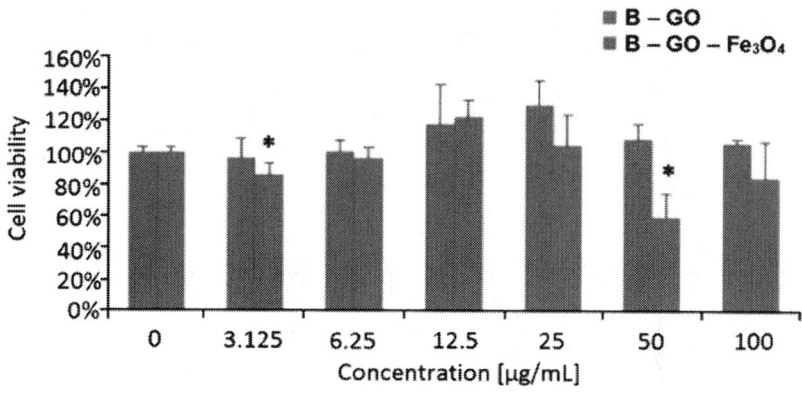

Figure 9: Relative viability of fibroblast cell line L929 exposed to B-GO and B-GO-Fe$_3$O$_4$hybrid.The cell viability is presented as percentage of the mean value. Bars represent standard deviation, and the symbol asterisk indicates statistically significant difference (p < 0.05).

Figure 10 presents cell viability upon interaction with N-GO-Fe$_3$O$_4$. Here, the best biocompatibility was noticed for the samples at concentration of 12.5 µg/mL for N-GO-Fe$_3$O$_4$ and N-GO, respectively. These results are similar to that obtained for samples with broader size distribution. However, at a dose of 100 µg/mL, the highest reduction of the cell viability was noticed. Little reduction of cell viability was noted at a dose 3.125 µg/mL in N-GO-Fe3O4 and GO sample. However, it can be observed that N-GO induces higher reduction of mitochondrial activity than hybrid samples. The same trend was observed in the sample with broad flake size distribution. It proves that the nanocomposite material is more biocompatible than pristine GO platforms. Furthermore, the chemical functionalization of GO and Fe_3O_4 leads to enhancement of the biocompatibility of the system and its independence of the size of GO. In the other study [35], functionalized GO and GO-Fe_3O_4 were tested on HeLa cell line. The WST-1 assay showed differences at mitochondrial activity between GO and GO-Fe_3O_4. Yang et al. indicated that the GO samples presented higher cytotoxicity than GO-Fe_3O_4[35]. Those results are in agreement with our study.

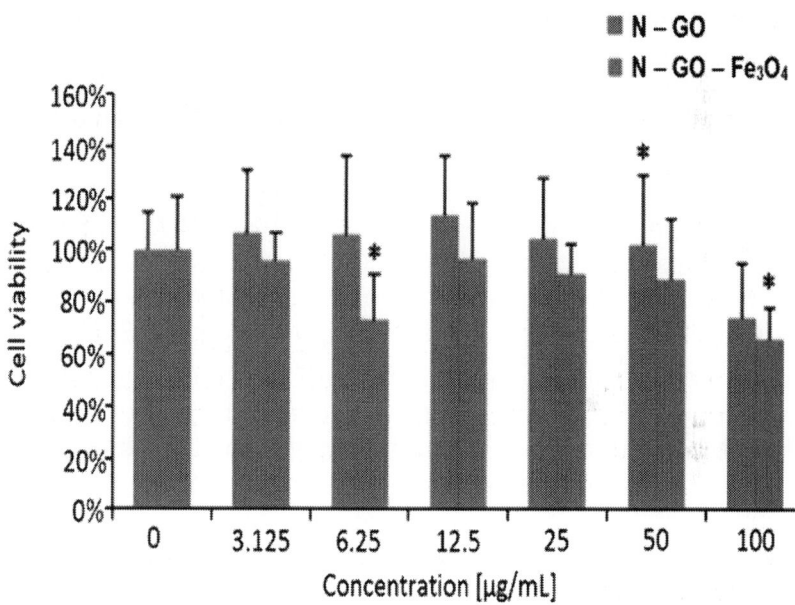

Figure 10: Relative viability of fibroblast cell line L929 exposed to N-GO and N-GO-Fe_3O_4 hybrid. The cell viability is presented as percentage of the mean value. Bars represent standard deviation. Symbol asterisk indicates statistically significant difference ($p < 0.05$).

The effects of GO on mouse fibroblast cells depend on GO dose, and as shown in the study of Wang et al. [55], the effects also depend on the culture time. The most cytotoxic effect of graphene oxide on human fibroblast cells (HDF) was observed on the fifth day of culture at the doses of 50 and 100 μg/mL. Similar results were noticed in tumor cell lines, e.g., human gastric cancer MGC803, human breast cancer MCF-7 and MDA-MB-435, and liver cancer HepG2 [55]. In our study, the effects of experimental samples on the cell culture were monitored for 24-h period, but as mentioned earlier, cell viability was reduced the most, to approximately 60%, at GO›s concentration of 50 µg/mL. Chang et al. [56] using CCK-8 assay and A549 cells made observation that preparation method of GO has influence of relative cell viability. The influence of different GO samples (s-GO with smaller size, l-GO with larger size, and m-GO mix) on the mitochondria activity may vary. m-GO›s effect on cell cultures was insignificant at the concentration range of 100 to 200 µg/mL. When the s-GO was tested, the cell viability was reduced the most at concentration between 50 and 200 µg/mL. It also has been noticed that the difference between some studies might come from the different sample properties and various cell lines. Incubation time can also influence the cell response [55,57]. In our study, the GO sample with concentration of 100 µg/mL demonstrated weak toxicity. We suggest that higher concentration (100 µg/mL) of graphene oxide may influence on harder GO migration to cell cytoplasm. We also found that GO material shows relatively good cytocompatibility at the concentration of 12.5 µg/mL and that result corresponds to the result obtained by Wojtoniszak et al. [58].

For the Fe_3O_4 material, lower mitochondrial activity was noticed at concentration of 3.125 and 50 µg/mL (Figure 11). In the samples of Fe_3O_4 with concentration of 6.25 µg/mL, the viability was reduced to 90%. At a dose of 100 µg/mL, the highest mitochondrial activity was observed for the Fe_3O_4 material. Analysing the results demonstrated by Shundo et al., one can see that even higher concentration of iron oxide nanoparticles, between 125 and 1,000 µg/mL, does not reduce the cell viability [59]. This can be explained by low uptake of Fe_3O_4 samples by the cells. Our study suggests that through the deposition of iron oxide nanoparticles via covalent linkage on the graphene oxide platform, the uptake of the nanomaterials is enhanced because the cell viability was affected more significantly than the pristine iron oxide nanoparticles. Little toxic effect of Fe_3O_4 on Cos-7 monkey kidney cells and GH3 pituitary tumor cells was observed in other analysis[60,61]. Figure 11 showed no dose-response relationship. Studies performed by Kai et al. indicated that the highest viability of BEL-7402 human hepatoma was observed when Fe_3O_4 nanoparticles at the concentration of 0.05 mg/mL were introduced to the cell culture [62]. When MgNPs-Fe_3O_4 was tested on A549 cell line for 24 h, no change in the cell viability was observed. The results of the Alamar Blue assay showed that treatment with 100 µg/mL of MgNPs-Fe$_3$O$_4$ for 72 h caused a significant reduction of cell viability [63].

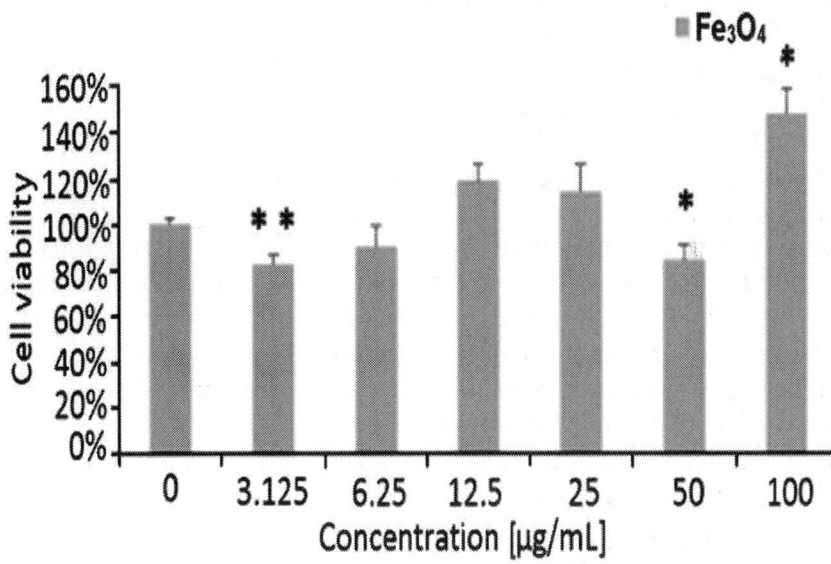

Figure 11: Relative viability of fibroblast cell line L929 exposed to Fe$_3$O$_4$ nanoparticles. The cell viability is presented as percentage of the mean value. Bars represent standard deviation, and symbol asterisk indicates statistically significant difference ($p < 0.05$).

As shown above, some difficulties in the interpretation of the obtained results can arise from variety of the factors that can influence the cell response [63]. Some of the factors are not clearly determined. Regarding the effects of graphene oxide and hybrid GO-Fe$_3$O$_4$ on cell viability, the mechanism is not well explained and still requires further analysis.

The in vitro studies play key role in exploration of the nanomaterial properties in biological environment and interaction with the living matter. The toxicity of the magnetic nanoparticles on biological entities is highly dependent on a range and combination of factors related to the properties of those nanoparticles. The physical properties such as the particle size, shape, and surface coating can evoke a toxic response by aggregating and coagulating according to size and shape. The chemical composition of the particles themselves can be naturally toxic. Here, we clearly demonstrate that the chemical functionalization of GO and Fe_3O_4 is a way to enhance the biocompatibility of the system and makes the system independent of the size of graphene oxide. Therefore, we believe that the obtained product with high cytocompatibility would be suitable for the application in biomedicine, e.g., as a drug carrier and/or in hyperthermia.

CONCLUSIONS

We report a facile method of the preparation of graphene oxide-Fe_3O_4 nanoparticle hybrid. We prove that it is possible to increase biocompatibility of graphene oxide through the deposition of magnetite nanoparticles on the graphene oxide flakes via chemical interaction. Furthermore, we indicate that the differences in flake size do not result in different cell viability in contact with our systems. These results show the potential application of this hybrid in hyperthermia treatment. Further investigation needs to be performed in order to prove the safety and efficiency of these systems in vivo.

AUTHORS' CONTRIBUTIONS

KU and MA carried out the synthesis and characterization of graphene oxide and graphene oxide-magnetite nanoparticle nanocomposites. RR carried out the synthesis of magnetite nanoparticles. MJ and MJ participated in the cytocompatibility studies and performed statistical analysis. EM and XC participated in the design of the study and coordination and helped to draft the manuscript. All authors read and approved the final manuscript.

ACKNOWLEDGEMENTS

This research was funded by the National Science Center under OPUS Program (Project No. DEC/2011/03/B/ST5/03239).

REFERENCES

1. Park S, Ruoff RS: Chemical methods for the production of graphenes. Nat Nanotechnol 2009, 4:217-224.

2. Rao CNR, Sood AK, Subrahmanyam KS, Govindraj A: Graphene: the new two-dimensional nanomaterial. Angew Chem Int Ed 2009, 48:7752-7777.

3. Geim AK: Graphene: status and prospects. Science 2009, 324:1530-1534.

4. Geim AK, Novoselov KS: The rise of graphene. Nat Mater 2007, 6:183-191.

5. Katsnelson MI: Graphene: carbon in two dimensions. Mater Today 2007, 10:20-27.

6. Loh KP, Bao Q, Ang PK, Yang J: The chemistry of graphene. J Mater Chem 2010, 20:2277-2289.

7. Loh KP, Bao Q, Eda G, Chhowalla M: Graphene oxide as a chemically tunable platform for optical applications. Nat Chem 2010, 2:1015-1024.

8. Nakada K, Fujita M, Dresselhaus G: Edge state in graphene ribbons: nanometer size effect and edge shape dependence. Phys Rev B 1996, 54:17954-17961.

9. He H, Klinowski J, Forster M: A new structural model for graphite oxide. Chem Phys Lett 1998, 287:53-56.

10. Lerf A, He H, Forster M: Structure of graphite oxide revisited. J Phys Chem B 1998, 102:4477-4482.

11. Kassaee MZ, Motamedi E, Majdi M: Magnetic Fe_3O_4-graphene oxide/polystyrene: Fabrication and characterization of a promising nanocomposite. Chem Eng J 2011, 172:540-549.

12. Yu MK, Jeong YY, Park J, Park S, Kim JW, Min JJ, Kim K, Jon S: Drug-loaded superparamagnetic iron oxide nanoparticles for combined cancer imaging and therapy in vivo. Angew Chem Int Ed Engl 2008, 47:5362-5365.

13. Larsen EKU, Nielsen T, Wittenborn T, Birkedal H, Vorup-Jensen T, Jacobsen MH, Ostergaard L, Horsman MR, Basenbacher F, Howard KA, Kjems J: Size-dependent accumulation of PEGylated silane-coated magnetic iron oxide nanoparticles in murine tumors. ACS Nano 2009, 3:1947-1951.

14. Miller MM, Prinz GA, Cheng SF, Bounnak S: Detection of a micron-sized magnetic sphere using a ring-shaped anisotropic magnetoresistance-based sensor: a model for magnetoresistance-based biosensor. Appl Phys Lett 2002, 81:2211-2213.

15. Jain TK, Morales MA, Sahoo SK, Leslie –Pelecky DL, Labhasetwar V: Iron oxide nanoparticles for sustained delivery of anticancer agents. Mol Pharm 2005, 2:194-205.

16. Chourpa L, Douziech–Eyrolles L, Ngaboni–Okassa L, Fouquenet JF, Cohen-Jonathan S, Souce M, Marchais H, Dubois P: Molecular composition of iron oxide nanoparticles, precursors for magnetic drug targeting, as characterized by confocal Raman microspectroscopy. Analyst 2005, 130:1395-1403.

17. Bulte JW: Intracellular endosomal magnetic labeling of cells. Methods Mol Med 2006, 124:419-439.

18. Modo M, Bulte JW: Cellular MR imaging. Mol Imagine 2005, 4:143-164.

19. Mornet S, Vasseur S, Grasset F, Duguet E: Magnetic nanoparticle design for medical diagnosis and therapy. J Mater Chem 2004, 14:2161-2175.

20. Dobson J: Magnetic nanoparticles for drug delivery. Drug Dev Res 2006, 67:55-60.

21. Derfus A, von Maltzahn G, Harris TJ, Duza T, Vecchio KS, Ruoslahti E, Bhatia SN: Remotely triggered release from magnetic nanoparticles. Adv Mater 2007, 19:3932-3936.

22. Jordan A, Scholz R, Maier–Hauf K, Johannsen M, Wust P, Nadobny J, Schirra H, Schmidt H, Deger S, Loening S, Lanksch W, Felix R: Presentation of a new magnetic field therapy system for the treatment of human solid tumors with magnetic fluid hyperthermia. J Magn Magn Mater 2001, 225:118-126.

23. Huh Y–M, Jun Y–W, Song H–T, Kim S, Choi J-S, Lee J-H, Yoon S, Kim K-S, Shin J-S, Suh J-S, Cheon J: In vivo magnetic resonance detection of cancer by using multifunctional magnetic nanocrystals. J Am Chem Soc 2005, 127:12387-12391.

24. Moroz P, Jones SK, Gray BN: Magnetically mediated hyperthermia: current status and future directions. Int J Hyperthermia 2002, 18:267-284.

25. Jordan A, Wust P, Fahling H, John W, Hinz A, Felix R: Inductive heating of ferromagnetic particles and magnetic fluids: physical evaluation of their potential for hyperthermia. Int J Hyperthermia 1993, 9:51-68.

26. Gupta AK, Gupta M: Synthesis and surface engineering of iron oxide nanoparticles for biomedical applications. Biomater 2005, 26:3995-4021.

27. Mahmoudi M, Milani AS, Stroeve P: Surface architecture of superparamagnetic iron oxide nanoparticles for application in drug delivery and their biological response: a review. Int J Biomed Nanosci Nanotechnol 2010, 1:164-201.

28. Hu Y, Li F, Bai X, Li D, Hua S, Wang K, Niu L: Label-free electrochemical impedance sensing of DNA hybridization based on functionalized graphene sheets. Chem Commun 2011, 47:1743-1745.

29. Hu K, Liu J, Chen J, Huang Y, Zhao S, Tian J, Zhang G: An amplified graphene oxide-based fluorescence aptasensor based on target-triggered aptamer hairpin switch and strand-displacement polymerization recycling for bioassays. Biosens Bioelectron 2013, 42:598-602.

30. Liu M, Zhao H, Quan X, Chen S, Fan X: Distance-independent quenching of quantum dots by nanoscale graphene in self-assembled sandwich immunoassay. Chem Commun 2010, 46:7909-7911.

31. Guo CX, Zheng XT, Lu ZS, Lou XW, Li CM: Biointerface by cell growth on layered graphene-artificial peroxidase-protein nanostructure for in situ quantitative molecular detection. Adv Mater 2010, 22:5164-5167.

32. Hu P, Zhu C, Jin L, Dong S: An ultrasensitive fluorescent aptasensor for adenosine detection based on exonuclease III assisted signal amplification. Biosens Bioelectron 2012, 34:83-87.

33. Feng B, Guo L, Wang L, Li F, Lu J, Gao J, Fan C, Huang Q: A graphene oxide-based fluorescent biosensor for the analysis of peptide-receptor interactions and imaging in somatostatin receptor subtype 2 overexpressed tumor cells. Anal Chem 2013, 85:7732-7737.

34.   Sun X, Liu Z, Welsher K, Robinson JT, Goodwin A, Zaric S, Dai H: Nano-graphene oxide for cellular imaging and drug delivery. Nano Res 2008, 1:203-212.

35.   Yang XY, Wang YS, Huang X, Ma Y, Huang Y, Yang R, Duan H: Multi-functionalized graphene oxide based anticancer drug-carrier with dual-targeting function and pH-sensitivity. J Mater Chem 2011, 21:3448-3454.

36.   Liu Z, Robinson JT, Sun XM, Dai HJ: PEGylated nanographene oxide for delivery of water-insoluble cancer drugs. J Am Chem Soc 2008, 130:10876-10877.

37.   Zhang LM, Lu ZX, Zhao QH, Huang J, Shen H, Zhang Z: Enhanced chemotherapy efficacy by sequential delivery of siRNA and anticancer drugs using PEI-grafted graphene oxide. Small 2011, 7:460-464.

38.   Zhang YB, Ali SF, Dervishi E, Xu Y, Li Z, Casciano D, Biris AS: Cytotoxicity effects of graphene and single-wall carbon nanotubes in neural phaeochromocytoma-derived PC12 cells. ACS Nano 2010, 8:3181-3186.

39.   Wang H, Shi L, Yan T, Zhang J, Zhong Q, Zhang D: Design of graphene-coated hollow mesoporous carbon spheres as high performance electrodes for capacitive deionization. J Mater Chem A 2014, 2:4739-4750.

40.   Wen X, Zhang D, Yan T, Zhang J, Shi L: Three-dimensional graphene-based hierarchically porous carbon composites prepared by a dual-template strategy for capacitive deionization. J Mater Chem A 2013, 1:12334-12344.

41.   Cong HP, He JJ, Lu Y, Yu SH: Water – soluble magnetic-functionalized reduced graphene oxide sheets: in situ synthesis and magnetic resonance imaging applications. Small 2010, 6:169-173.

42.   Shen X, Wu J, Bai S, Zhou H: One-pot solvothermal syntheses and magnetic properties of graphene-based magnetic nanocomposites. J Alloys Compd 2010, 506:136-140.

43.   He H, Gao C: Supraparamagnetic, conductive, and processable multifunctional graphene nanosheets coated with high-density Fe_3O_4 nanoparticles. ACS Appl Mater Interfaces 2010, 2:3201-3210.

44.   He F, Fan J, Ma D, Zhang L, Leung C, Chan HL: The attachment of Fe_3O_4 nanoparticles to graphene oxide by covalent bonding. Carbon 2010, 48:3139-3144.

45.   Zheng XT, Li CM: Restoring basal planes of graphene oxides for highly efficient loading and delivery of beta-lapachone. Mol Pharm 2012, 9:615-621.

46.   Bai L–Z, Zhao D–L, Xu Y, Zhang J-M, Gao Y-L, Zhao L-Y, Tang J-T: Inductive heating property of graphene oxide-Fe_3O_4 nanoparticles hybrid in an AC magnetic field for localized hyperthermia. Mater Lett 2012, 68:399-401.

47.   Marcano DC, Kosynkin DV, Berlin JM, Sinitskii A, Sun Z, Slesarev A, Alemany LB, Lu W, Tour JM: Improved synthesis of graphene oxide. ACS Nano 2010, 4:4806-4814.

48.   Rakoczy R: Mixing energy investigations in a liquid vessel that is mixed by using a rotating magnetic field. Chemical Eng Process 2013, 66:1-11.

49.   Ishiyama M, Tominaga H, Shiga M, Sasamoto K, Ohkura Y, Ueno K, Watanabe M: Novel cell proliferation and cytotoxicity assays using a tetrazolium salt that produces a water-soluble formazan dye. In Vitro Toxicol 1995, 8:187-190.

50.   Johnsen AR, Bendixen K, Karlson U: Detection of microbial growth on polycyclic aromatic hydrocarbons in microtiter plates by using the respiration indicator WST-1. Appl Environ Microbiol 2002, 68:2683-2689.

51.   Berridge MV, Tan AS: High-capacity redox control at the plasma membrane of mammalian cells: trans-membrane, cell surface, and serum NADH-oxidases. Antioxid Redox Signal 2000, 2:231-242.

52.   Langford JI, Wilson AJC: Scherrer after sixty years: a survey and some new results in the determination of crystallite size. J Appl Cryst 1978, 11:102-113.

53.   Ferrari AC, Meyer JC, Scardaci V, Casiraghi C, Lazzeri M, Mauri F, Piscanec S, Jiang D, Novoselov KS, Roth S, Geim AK: Raman spectrum of graphene and graphene layers. Phys Rev Lett 2006, 97:187401.

54.   Ferrari AC: Raman spectroscopy of graphene and graphite: disorder, electron-phonon coupling, doping and nonadiabatic effects. Solid State Commun 2007, 143:47-57.

55. Wang K, Ruan J, Song H, Zhang J, Wo Y, Guo S, Cui D: Biocompatibility of graphene oxide. Nano Res Lett 2011, 6:1-8.

56. Chang Y, Yang ST, Liu JH, Dong E, Wang Y, Cao A, Liu Y, Wang H: In vitro toxicity evaluation of graphene oxide on A549 cells. Toxic Lett 2011, 200:201-210.

57. Lim HN, Huang NM, Lim SS, Harrison I, Chia CH: Fabrication and characterization of graphene hydrogel via hydrothermal approach as a scaffold for preliminary study of cell growth. Inter J Nanomed 2011, 6:1817-1823.

58. Wojtoniszak M, Chen X, Kalenczuk RJ, Wajda A, Lapczuk J, Kurzewski M, Drozdzik M, Chu PK, Borowiak–Palen E: Synthesis, dispersion, and cytocompatibility of graphene oxide and reduced graphene oxide. Colloids Surf B: Biointerfaces 2012, 89:79-85.

59. Shundo C, Zhang H, Nakanishi T, Osaka T: Cytotoxicity evaluation of magnetite (Fe_3O_4) nanoparticles in mouse embryonic stem cells. Coll Surf B: Bionterfaces 2012, 97:221-225.

60. Shieh DB, Cheng FY, Su CH, Yeh CS, Wu MT, Wu YN, Tsai CY, Wu CL, Chen DH, Chou CH:Aqueous dispersion of magnetite nanoparticles with NH_3^+ surfaces for magnetic manipulations of biomolecules and MRI contrast agents. Biomat 2005, 26:7183-7191.

61. Liu YC, Wu PC, Shieh DB, Wu SN: The effects of magnetite (Fe_3O_4) nanoparticles on electroporation-induced inward currents in pituitary tumor (GH_3) and in RAW 264.7 macrophages. Int J Nanomedicine 2012, 7:168-1696.

62. Kai W, Xiaojun X, Ximing P, Zhenqing H, Qiqing Z: Cytotoxic effects and the mechanism of three types of magnetic nanoparticles on human hepatoma BEL-7402 cells. Nano Res Lett 2011, 6:480-490.

63. Watanabe M, Yoneda M, Morohashi A, Hori Y, Okamoto D, Sato A, Kurioka D, Nittami T, Hirokawa Y, Shiraizhi T, Kawai K, Kasai H, Totsuka Y: Effects of Fe_3O_4 magnetic nanoparticles on A549 cells. Int J Mol Sci 2013, 14:15546-15560.

Chapter 4

Hydrothermal-Assisted Exfoliation of Y/Tb/Eu Ternary Layered Rare-Earth Hydroxides into Tens of Micron-Sized Unilamellar Nanosheets for Highly Oriented and Color-Tunable Nano-Phosphor Films

Qi Zhu[1], Zhixin Xu[1], Ji-Guang Li[1,2], Xiaodong Li[1,] Yang Qi[3], and Xudong Sun[1]

[1]Key Laboratory for Anisotropy and Texture of Materials (Ministry of Education), School of Materials and Metallurgy, Northeastern University, No. 3-11, Wenhua Road, Shenyang 110819, Liaoning, China

[2]Advanced Materials Processing Unit, National Institute for Materials Science, Namiki 1-1, Tsukuba 305-0044, Ibaraki, Japan

[3]Institute of Materials Physics and Chemistry, School of Sciences, Northeastern University, No. 3-11, Wenhua Road, Shenyang 110819, Liaoning, China

ABSTRACT

Efficient exfoliation of well-crystallized $(Y_{0.96}Tb_xEu_{0.04-x})_2(OH)_5NO_3 \cdot nH_2O$ $(0 \leq x \leq 0.04)$ layered rare-earth hydroxide (LRH) crystals into tens of micron-sized unilamellar nanosheets has been successfully achieved by inserting water insoluble oleate anions $(C_{17}H_{33}COO^-)$ into the interlayer of the LRH via hydrothermal anion exchange at 120°C, followed by delaminating in toluene. The intercalation of oleate anions led to extremely expanded interlayer distances (up to approximately 5.2 nm) of the LRH crystals and accordingly disordered stacking of the ab planes along the c-axis and also weakened interlayer interactions, without significantly damaging the ab plane. As a consequence, the thickness of the LRH crystals increased from approximately 1 to 10 μm, exhibiting a behavior similar to that observed from the smectite clay in water. Highly [111]-oriented and approximately 100-nm thick oxide films of $(Y_{0.96}Tb_xEu_{0.04-x})_2O_3$ $(0 \leq x \leq 0.04)$ have been obtained through spin-coating of the exfoliated colloidal nanosheets on quartz substrate, followed by annealing at 800°C. Upon UV excitation at 266 nm, the oxide transparent films exhibit bright luminescence, with the color-tunable emission from red to orange, yellow, and then green by increasing the Tb^{3+} content from $x = 0$ to 0.04.

BACKGROUND

Layered inorganic compounds have interesting physical/chemical properties, such as tunable interlayer spacing and interlayer composition, and can be readily functionalized via intercalation to produce specific properties [1]. In addition, they may potentially be delaminated into unilamellar Nano sheets or Nano sheets of few-layer thick via ion exchange, followed by mechanical agitation in a proper medium [1]. The obtained nanosheets can serve as ideal building blocks for the construction of inorganic or hybrid organic–inorganic multifunctional films owing to their significantly two-dimensional morphologies (lateral size up to microns and thickness down to nanometer level) [2]-[6]. Because of the significant morphological anisotropy, the nanosheets tend to orient themselves, with a certain crystallographic direction perpendicular to substrate surface, and thus introduce additional or greatly enhanced functionalities. Delaminating layered compounds into nanosheets attracted much attention, and monolayer nanosheets have been successfully exfoliated from several types of layered inorganic materials, such as layered double hydroxides (LDHs) [7],[8], graphite [9], metal oxides[10], phosphates [11], and chalcogenides [12].

Layered rare-earth hydroxides (LRHs) [13]-[26], with a general formula of $RE_2(OH)_5(A^{m-})_{1/m} \cdot nH_2O$ (rare-earth (RE) ions; intercalated (A) anions), are a new group of important anion-type layered materials that may potentially be exfoliated into single or few-layer thick nanosheets for the further construction of various nanostructures, particularly transparent functional films. Due to the unique electronic, optical, magnetic, and catalytic properties of the rare-earth elements, LRHs attracted immediate attentions for controllable synthesis since their emergence, and some efforts have been paid to the thinning of LRHs via exfoliation [22], [23],[26] and exfoliation-free synthesis [24]. Recently, LRHs crystals have been exfoliated into nanosheets by several research groups via anion exchange with dodecylsulfate (DS⁻) at room temperature, followed by mechanical agitation in formamide [22],[23]. Despite these successes, the high-charge density of LRHs makes a complete exfoliation rather challenging. Previous studies also showed that exfoliation usually takes several days and is thus an arduous and lengthy work [22]. We have obtained in our previous work ultra-thin LRHs nanosheets (down to approximately 4 nm), without exfoliation, by capping thickness growth of the crystals with tetrabutylammonium ions (TBA⁺, $(C_4H_9)_4N^+$) in hydrothermal reaction, but the nanosheets are limited to submicron in lateral size [24]. It is also worth noting that the LRHs particles synthesized through the current techniques are mostly platy crystals of several micrometers in lateral dimension, so the final exfoliated nanosheets are submicron sized [22]. Very recently, unilamellar nanosheets with lateral sizes ≥60 μm and thicknesses of only approximately 1.6 nm have been efficiently delaminated by us from sub-millimeter-sized LRHs crystals [25] via hydrothermal anion exchange of the interlayer NO_3^- with dodecylsulfate ($C_{12}H_{25}OSO_3^-$,

DS$^-$), followed by exfoliation in formamide. Significantly faster anion exchange and higher extent of DS$^-$ intercalation were observed for the hydrothermal than ambient processing [26].

The interlayer space of LRH is significantly affected by the size of the intercalated anions, and a more weakened interlayer interaction via insertion of bigger anions is beneficial to exfoliation. Water-soluble dodecylsulfate ($C_{12}H_{25}OSO_3^-$, DS$^-$), which has a long carbon chain, is usually employed to swell anion-type layered compounds for delamination via room temperature anion exchange, and successes were manifested in the cases of LDHs [7],[8] and LRHs [22]. Anions of even longer carbon chain, such as oleate ($C_{17}H_{33}COO^-$), would be more efficient for interlayer expansion but are hardly soluble in water at room temperature. We show in this work the successful insertion of oleate anions into the interlayers of tens of micron-sized LRH crystals via hydrothermal anion exchange and based on which the efficient exfoliation of ultra-large (approximately 20 µm) and single layer (approximately 1.55 nm) nanosheets in toluene. Highly [111]-oriented oxide films have also been constructed via self-assembly of the resultant nanosheets for multi-color emissions. In the following sections, we report the hydrothermal intercalation of oleate into LRHs crystals of the Y/Tb/Eu ternary system, exfoliation of nanosheets, and assembly of transparent films with the nanosheets for color-tunable emissions. The materials are characterized in detail by the combined techniques of field emission scanning electron microscopy (FE-SEM), transmission electron microscopy (TEM), X-ray diffraction (XRD), Fourier transform infrared spectroscopy (FTIR), atomic force microscopy (AFM), and optical spectroscopy, and we believe that the outcomes of this work would have wide implications to other layered inorganic materials.

METHODS

Synthesis

The starting rare-earth sources for LRH synthesis are Y_2O_3, Tb_4O_7, and Eu_2O_3, all 99.99% pure products from Huizhou Ruier Rare-Chem. Hi-Tech. Co. Ltd (Huizhou, China). Analytical grade nitric acid (HNO_3, 63 wt.%), ammonium hydroxide solution ($NH_3 \cdot H_2O$, 25 wt.%), and ammonium nitrate (NH_4NO_3, 99.0% pure) were purchased from Shenyang Chemical Regent Factory (Shenyang, China). The nitrate solution of RE^{3+} was prepared by dissolving the corresponding oxide with a slightly excessive amount of nitric acid, followed by evaporation at approximately 90°C to dryness to remove the superfluous acid. Synthesis of ultra-large LRH crystals for the Y/Tb/Eu ternary system was conducted via hydrothermal reaction (180°C for 24 h) in the presence of NH_4NO_3, as described in our previous paper [25]. The optimal concentration of either Tb^{3+} or Eu^{3+} in Y_2O_3 is approximately 4 to 5 at.%, above which concentration quenching of luminescence would take place. This value would hold for the Tb^{3+}/Eu^{3+} pair, and thus the total concentration of Tb^{3+} and Eu^{3+} is fixed at 4 at.% in this work.

Hydrothermal-assisted Anion Exchange, Exfoliation, and Film Construction

For a typical anion-exchange reaction, 0.4 mmol of $(Y_{0.96}Tb_xEu_{0.04-x})_2$ $(OH)_5NO_3 \cdot nH_2O$ ($0 \leq x \leq 0.04$) was dispersed in 50 mL of water containing a proper amount of sodium oleate ($C_{17}H_{33}COONa$). The resultant suspension was transferred into a Teflon lined stainless-steel autoclave of 100 mL capacity after being stirred for 5 min. The autoclave was tightly sealed and was put in an electric oven preheated to 120°C. After 24 h of reaction, the autoclave was left to cool naturally to room temperature, and the anion-exchange product was collected via centrifugation. The wet precipitate was washed with hot distilled water (80°C) for three times, rinsed with absolute ethanol, and was finally dried in air at 50°C for 24 h. The anion-exchange product was then dispersed in 50 mL of toluene, and a transparent colloidal suspension was obtained after constant magnetic stirring for 12 h. The resultant nanosheets were assembled into films on quartz substrates (10 mm in diameter) via spin-coating. Briefly, 200 μL of the transparent colloidal suspension was dropped on the substrate fixed on a spin coater, spun at 2,000 revolutions per minute (rpm) for 1 min to assemble the nanosheets, followed by slow air drying. Prior to spin coating, the quartz substrate was ultrasonically cleaned in sequence in acetone, ethanol, and distilled water, and was then immersed in a mixed solution (3:1 in volume ratio) of H_2SO_4 (30 vol%) and H_2O_2 (30 vol.%) heated to 80°C for 1 h. Subsequently, the substrate was kept in the mixed solution of $H_2O:NH_4OH:30$ vol% H_2O_2 (5:1:1 in volume ratio) to render surface hydrophilicity. Before use, the substrate was immersed in an aqueous solution of polyethylenimine (PEI, 1.5 mg/mL), soaked in a poly sodium 4-styrene sulfonate (PSS) aqueous solution (1 mg/mL) for 1 h, followed by washing with distilled water for three times and drying. Oxide film was obtained by calcining the LRH film in flowing oxygen (200 mL/min) at 800°C for 4 h, followed by reducing in flowing hydrogen (200 mL/min) at 800°C for 2 h for the Tb^{3+}containing samples.

Characterization Techniques

Phase identification was performed by XRD (Model PW3040/60, Philips, Eindhoven, the Netherlands) operating at 40 kV/40 mA using nickel-filtered Cu K radiation and a scanning speed of 4.0° 2/min. Lattice constants were calculated from the XRD patterns using the software package X'Pert HighScore Plus version 2.0 (PANanalytical B.V., Almelo, the Netherlands). Morphologies of the products were observed via FE-SEM (Model JSM-7001 F, JEOL, Tokyo, Japan) and TEM (Model JEM-2000FX, JEOL, Tokyo). FTIR (Model Spectrum RXI, Perkin-Elmer, Shelton, CT, USA) of the pristine and anion-exchanged LRHs was performed by the standard KBr method. Chemical composition of the products was determined via elemental analysis for the Y/Tb/Eu content by the inductively coupled plasma spectrophotometric method with an accuracy of 0.01 wt.% (ICP, Model IRIS Advantage, Nippon Jarrell-Ash Co. Ltd., Kyoto, Japan), for NO_3^- via spectrophotometry (Ubest-35, Japan Spectroscopic Co., Ltd., Tokyo, Japan), and for the carbon content on a simultaneous carbon/sulfur determinator with a detection limit of 0.01 wt.% (Model CS-444LS, LECO, St. Joseph, MI, USA). A Nanosurf easyScan 2 AFM (Nanosurf, Liestal, Switzerland) was employed to obtain topographical images of the nanosheets. Optical properties of the phosphor films were measured at room temperature with a UV–vis spectrophotometer (Lambda-750S, Perkin-Elmer, Shelton, CT, USA) for transmittance and with an LS-55 fluorescence spectrophotometer (Perkin-Elmer, Shelton, CT, USA) for photoluminescence excitation (PLE) and emission (PL).

RESULTS AND DISCUSSION

Well-crystallized and ultra-large LRHs crystals can be synthesized by autoclaving mixed nitrate solution of the component rare-earths at 180°C to 200°C and in the presence of NH_4NO_3 mineralizer [25]. Figure 1a shows XRD patterns of the hydrothermal products with various amounts of Tb^{3+}. A series of strong (00 l) reflections were observed to be characteristic of a layered phase, as previously reported for the $Ln_2(OH)_5NO_3 \cdot nH_2O$ LRHs. A number of weak non-(00 l) reflections were also detected, indicating that the LRHs are ordered for their hydroxide main layers. The solid solutions were found to have similar lattice constants of $a \sim 1.273$, $b \sim 0.715$, and $c \sim 1.800$ nm, due to the small total content (4 at.%) of Tb^{3+} and Eu^{3+} and the similar sizes of

the two kinds of ions (for eightfold coordination, $r_{Tb^{3+}} = 0.1040$nm and

$r_{Eu^{3+}} = 0.1066$nm). Figure 1b shows FE-SEM morphology of the $x = 0.035$ sample, where the crystals were observed to be uniform hexagons of ≥30 μm in lateral size and approximately 1 μm in thickness. The straight and sharp crystal edges with intersection angles of approximately 120° may suggest high crystallinity of the platelets. Similar morphologies were observed for the other Y/Tb/Eu combinations and are thus not shown.

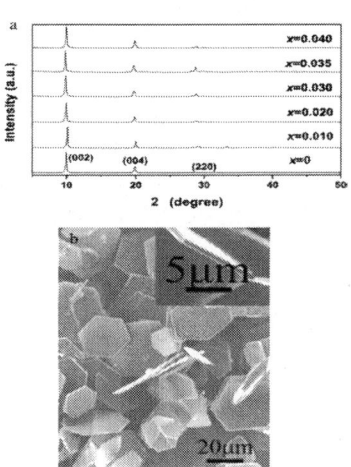

Figure 1: XRD patterns of the hydrothermal products and FE-SEM morphology of the x = 0.035 sample. (a) XRD patterns of the $(Y_{0.96}Tb_xEu_{0.04-}$

$_{x)_2}(OH)_5NO_3 \cdot nH_2O$ ($0 \leq x \leq 0.04$) LRH solid solutions hydrothermally synthesized at 180°C and (b) FE-SEM morphology of the $x = 0.035$ sample.

Figure 2 shows XRD patterns of the anion exchange products (denoted as LRH-oleate hereafter) obtained via hydrothermal reaction at 120°C for 24 h, from which it is seen that the non-(00 l) reflections, such as (220), disappeared from the LRH-oleate, though the (00 l) reflections are still observable. Several sets of (00 l) reflections were observed for each of the products, suggesting the existence of multi interlayer distances as found in our previous work [26]. For the $x = 0.040$ sample (Figure 2b), the sharp and symmetric (00 l) reflections at 1.734, 1.030, and 0.730 nm indicate a basal spacing of approximately 5.15 nm, those at 1.354 and 0.837 nm correspond to a basal spacing of approximately 4.10 nm, and the ones at 1.167 and 0.695 nm conform to a basal spacing of approximately 3.50 nm. Similarly, the $x = 0.030$ sample has basal spacings of approximately 5.20, 4.70, 4.20, and approximately 3.50 nm, and the $x = 0.00$ sample has spacing values of approximately 4.50 and 4.20 nm. The existence of multi-spacings suggests that the oleate anions may have moved into the interlayers of the LRHs via the 'wriggle intercalation' model proposed recently [26]. It was noted that the maximum basal spacing of approximately 5.20 nm is much larger than that of the LRH-DS$^-$ (approximately 3.70 nm) under identical hydrothermal anion-exchange [22], [26], revealing the higher efficiency of oleate over DS$^-$ in LRH swelling. In addition, quasi-amorphous diffractions were observed in the $2\vartheta \geq 15°$ range, different from the LRH-DS$^-$ obtained with the same hydrothermal treatment. In the latter case, immobile (220) and other non-(00 l) reflections are clearly observable [22], [26]. Since the oleate anion ($C_{17}H_{33}COO^-$) has a longer carbon chain than DS$^-$($C_{12}H_{25}OSO_3^-$), diffusion of a large amount of oleate into the interlayer gallery may have resulted in the higher degrees of swelling observed from Figure 2, similar to the 'osmotic hydration' of smectite clay in water and the interlayer expansion of LDH-DS$^-$ in formamide. The extremely expanded interlayer may disorder the stacking of ab planes in the [001] direction, and thus amorphous XRD diffractions appeared. The ab plane, however, was not significantly damaged, as inferred from the appearances of (00 l) reflections. The extremely swollen LRH-oleate with long-range ordered ab planes would be beneficial to exfoliation.

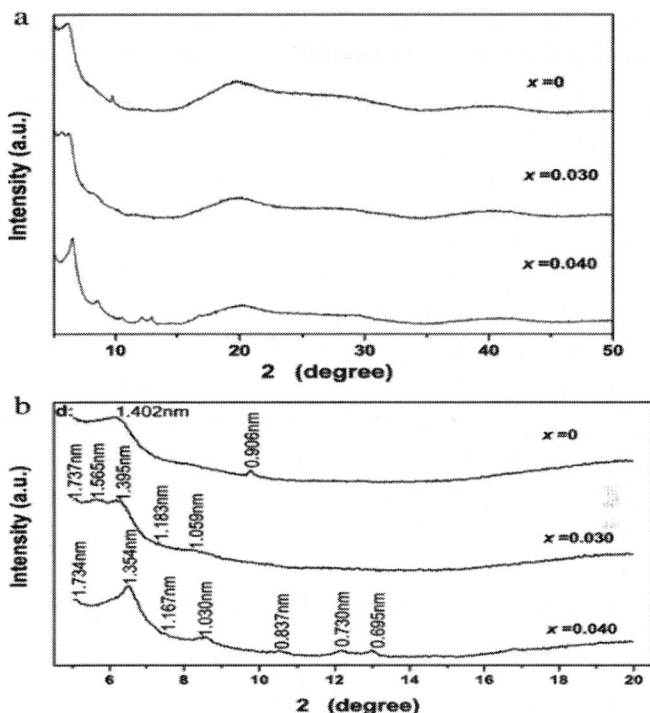

Figure 2: XRD patterns of the LRH-oleate samples obtained by hydrothermal anion exchange. (b) is part of (a) in the 2ϑ range of 4.5° to 20.5°.

Figure 3a shows FTIR spectra for the pristine LRH and LRH-oleate. For LRH, the absorption peaks at approximately 3,372 cm^{-1} and the shallow shoulder near 1,636 cm^{-1} provide evidence for water of hydration in the structure, and they are assignable to the O-H stretching vibrations (v_1 and v_3) and the H-O-H bending mode (v_2), respectively [27],[28]. The absorption band observed in the range of 3,500 to 3,750 cm^{-1} (centered at approximately 3,586 cm^{-1}) is indicative of hydroxyl (OH$^-$) groups [27], [28]. The strong absorption peak at 1,384 cm^{-1} is characteristic of an uncoordinated nitrate anion, as also found for other layered hydroxides containing free interlayer nitrate groups [13],[21],[27],[28]. After anion exchange, the vibration of nitrate is no longer observable. Instead, two intense bands appeared at approximately 1,572 and 1,454 cm^{-1}, which are assignable to the stretching modes of carboxyl (COO$^-$) [27], [28]. The strong absorptions at approximately 2,926 and 2,853 cm^{-1} are due to the asymmetric and symmetric CH_2 stretching vibrations, respectively, whereas the weak band at approximately 3,003 cm^{-1} is assignable to the stretching mode of the terminal CH_3 group of the hydrocarbon tail [27],[28]. The above results confirmed a complete replacement of the interlayer nitrate by oleate. Chemical analysis yielded a general formula of $Ln_2(OH)_5(C_{17}H_{33}COO)(C_{17}H_{33}COOH)y \cdot nH_2O$ (Ln = Y, Tb, Eu) by applying molecular neutrality, assuming that all the C are from $C_{17}H_{33}COO^-$ and $C_{17}H_{33}COOH$ The results of chemical analysis comply with the FTIR observations. Figure 3b shows the typical morphology of LRH-oleate, from which it is seen that the thickness of LRH platelets has been significantly expanded from approximately 1 to 10 µm, and, owing to the massive insertion of oleate anions, cracks of different gaps are formed along the thickness direction.

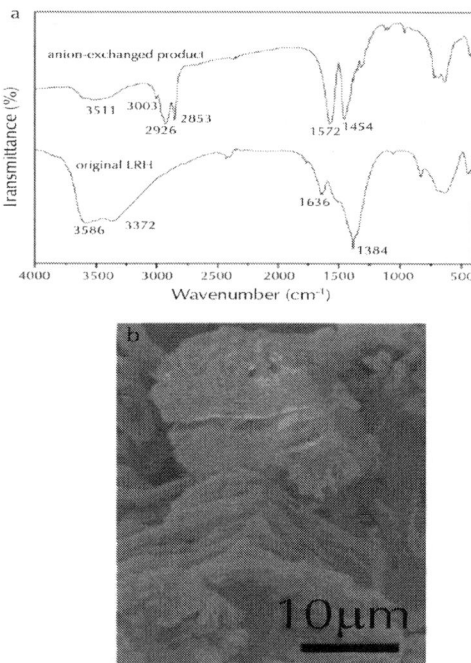

Figure 3: FTIR spectra for the pristine LRH and LRH-oleate and the typical morphology of LRH-oleate. (a) FTIR spectra for the pristine LRH (x = 0.035) and its oleate-exchange derivative. (b) FE-SEM morphology of the LRH-oleate sample. Similar results were observed for the rest samples studied in this work.

Dispersing the LRH-oleate in 50 mL of toluene yielded a transparent colloidal suspension via constant and slow magnetic stirring for 12 h. The clearly observable Tyndall effect under laser beam irradiation (inset in Figure 4a) indicates the delamination of LRH-oleate. FE-SEM observation found that most of the exfoliated nanosheets have lateral sizes of ≥20 μm (Figure 4a). The uniform contrast under TEM (Figure 4b) of the individual nanosheets implies that the nanosheet is rather thin. Selected area electron diffraction (SAED) yielded well-arranged spot-like patterns, suggesting that the nanosheet under observation is well crystallized and is of single crystalline (inset in Figure 4b). The cell parameters calculated from the SAED pattern are $a \sim 1.27$ and $b \sim 0.72$ nm, in good agreement with those of the bulk LRH [21]. The nanosheet was estimated to be approximately 1.55 nm thick from the AFM height profile (Figure 4e), indicating that the nanosheet is primarily of unilamellar. At the same time, AFM observation indicated that the nanosheets are very flat and smooth (Figures 4c,d). Possibly due to surface chemical adsorption of oleate and toluene molecules, the unilamellar nanosheets are a little thicker than the crystallographic thickness of 0.93 nm [24]. Compared with those reported previously [22],[23], the unilamellar nanosheets obtained in this work showed a significantly larger lateral size and a more unabridged shape, which is advantageous for the construction of highly oriented functional films.

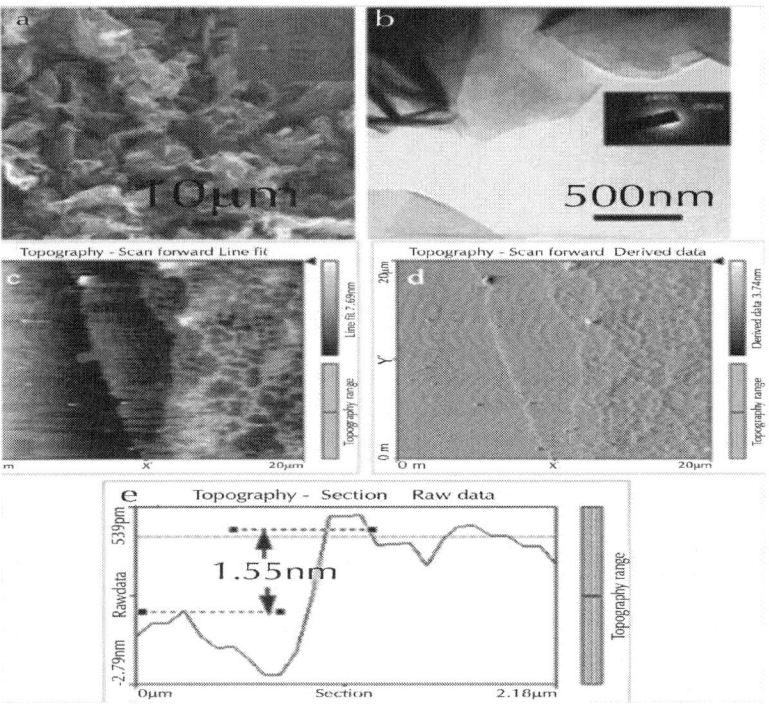

Figure 4: FE-SEM, TEM, AFM images and height profile of the exfoliated nanosheets. (a) FE-SEM and (b) TEM micrographs showing morphologies of the nanosheets exfoliated from LRH-oleate ($x = 0.035$). The AFM images (c,d) and the height profile (e) along the red line marked in(c), respectively. The inset in (a) shows the appearance of a colloidal suspension of the nanosheets in toluene, with a clearly observable Tyndall effect under laser beam irradiation. The inset in (b) is the SAED pattern of an individual unilamellar nanosheet.

Depositing 200 µL colloidal nanosheets (solid loading: approximately 2 vol.%) on a quartz substrate followed by spin-coating has produced highly c-axis-oriented films via self-assembly of the nanosheets (Figure 5a(g)). As there are oleate anions and toluene molecules on surfaces of the positively charged nanosheets, the nanosheets tend to assemble themselves into new layered materials similar to LRH, and thus the (00l) reflection was observed from the LRH film in Figure 5a(g). Calcining the LRH films at 800°C for 4 h, followed by hydrogen reduction at the same temperature for 2 h for the Tb-containing ones, yielded cubic-structured $(Y_{0.96}Tb_xEu_{0.04-x})_2O_3$ ($0 \leq x \leq 0.04$) films (Figure 5a(a-f)). Because the projection in the [001] direction for the LRHs crystal and in the [111] direction for the cubic oxide crystal present close similarities in terms of rare-earth atomic configuration, the phase transformation from LRH to oxide is a quasitopotactic one [24]. The oxide films are thus highly [111] oriented and show strong (222) while very weak non-(222) reflections. Calculation with the (222) diffraction yielded similar cell constants of approximately 1.0664 nm for all the oxide films, similarly due to the small total content of Tb^{3+} and Eu^{3+}. The oxide films are flat and significantly denser than those made with submicron-sized nanosheets [24], showing the great advantages of larger sheet size (Figure 5b). The oxide films were estimated to be approximately 100 nm thick via cross-section FE-SEM view (the inset in Figure 5b), and exhibit high transmittances of ≥75% (bare quartz: approximately 94%, Figure 5c) in the visible wavelength region (500 to 800 nm).

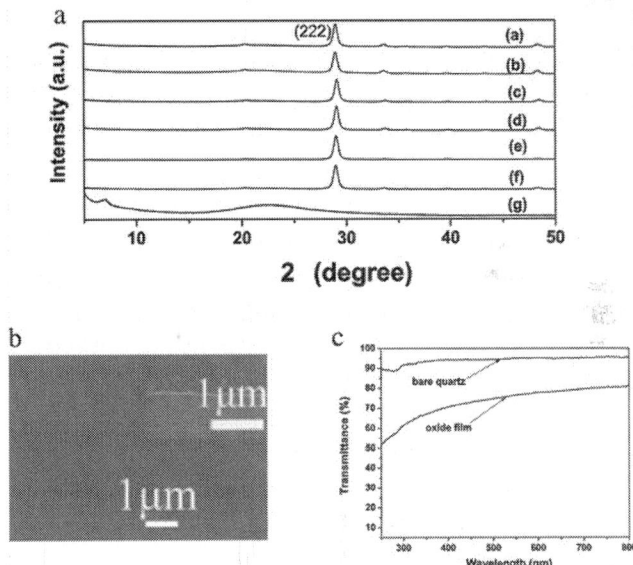

Figure 5: XRD patterns, FE-SEM image, and transmission spectrum of the films. (a) XRD patterns of the $(Y_{0.96}Tb_xEu_{0.04-x})_2O_3$ ($0 \leq x \leq 0.04$) oxide films calcined from the LRH films constructed with exfoliated nanosheets, where (a)-(f) correspond to $x = 0$, 0.01, 0.02, 0.03, 0.035, and 0.04, respectively. Line (g) in (a) is for the LRH nanosheet film ($x = 0.035$). (b) and (c) are FE-SEM image and transmission spectrum of the $(Y_{0.96}Tb_{0.035}Eu_{0.005})_2O_3$ film, respectively. The inset in (b) is the cross-section view of the film.

Although the macroscopic concentration of Tb^{3+}/Eu^{3+} activators is fixed at 4 at.% in this work, concentration difference exists among different crystal planes. The (222) facet of Y_2O_3 is a close-packed one and thus has higher Y^{3+} occupancy. As the activators randomly replace Y^{3+}, it can thus be said that more activators would reside on (222) to yield significantly enhanced emission. As shown in our previous work [24],[26], the [111]-oriented $(Y, Eu)_2O_3$ film exhibited an emission intensity ≥ 2 times that of the randomly oriented powder of the same composition. Therefore, the highly [111]-oriented $(Y_{0.96}Tb_xEu_{0.04-x})_2O_3$ $(0 \leq x \leq 0.04)$ films fabricated in this work are expected to yield bright emissions.

shows PLE/PL spectra of the highly [111]-oriented $(Y_{0.96}Eu_{0.04})_2O_3$ $(x = 0)$ and $(Y_{0.96}Tb_{0.04})_2O_3$ $(x = 0.04)$ films. For $(Y_{0.96}Eu_{0.04})_2O_3$, the excitation spectrum consists of a broad and intense band at around 240 nm, which can be assigned to the charge-transfer (CT) from O^{2-} to Eu^{3+}[29],[30]. Upon UV excitation at 240 nm, the oxide film exhibits sharp lines ranging from 500 to 700 nm, which are associated with the transitions from the excited $^5D_0$ to the ^7FJ $(J = 0,1,2,3)$ ground states of Eu^{3+}[29],[30]. Relative intensity of the transition to different J levels depends on the site symmetry of Eu^{3+}, and the dominant red emission at 613 nm arises from the hypersensitive $^5D_0 \rightarrow {}^7F_2$ forced electric dipole transition of Eu^{3+} taking the non-centrosymmetric C_2 lattice sites. The $(Y_{0.96}Tb_{0.04})_2O_3$ film exhibits a broad and strong excitation band in the 250- to 330-nm region with a maximum at 276 nm corresponding to the well-documented $4f^8 \rightarrow 4f^75d^1$ Tb^{3+} transition [31]. When excited at 276 nm, the oxide film displayed the typical $^5D_4 - {}^7FJ$ $(J = 5$ to $2)$ transitions of Tb^{3+} at about 543, 600, 627, and 671 nm, respectively, with the strongest emission being at 543 nm for green [31]. Overlapping the red- and green-emitting films yielded a bright yellow color under 254-nm radiation from a hand-held UV lamp, indicating that the emission color can be tuned by varying the Tb/Eu molar ratio. The excitation spectra of $(Y_{0.96}Eu_{0.04})_2O_3$ $(x = 0)$ and $(Y_{0.96}Tb_{0.04})_2O_3$ $(x = 0.04)$ intersect at 266 nm, which would thus represent the most efficient wavelength to simultaneously excite Eu^{3+} and Tb^{3+}.

Photoluminescence (PL) of the $(Y_{0.96}Tb_xEu_{0.04-x})_2O_3$ films $(0 \leq x \leq 0.04)$ were systematically investigated to define the emission behavior and emission color (Figure 6a). It is seen that the Eu^{3+} emission at approximately 613 nm monotonically decreases while the Tb^{3+} emission at 543 nm improves with increasing Tb incorporation

(Figures 6a and 7). We analyzed in Figure 7 the relative intensities of these two peaks together with the $I(^5D_0 \rightarrow {}^7F_2)/I(^5D_4 \rightarrow {}^7F_5)$ intensity ratio as a function of the Tb content ($0.010 \leq x \leq 0.035$). Clearly, the ratio decreases from approximately 2.0 at $x = 0.010$ to approximately 0.34 at $x = 0.035$, suggesting color-tunable emissions. When excited at 266 nm, the samples have Commission Internationable Ed l'eclairage (CIE) chromaticity coordinates of (0.56, 0.44) for $x = 0$, (0.50, 0.50) for $x = 0.010$, (0.47, 0.52) for $x = 0.020$, (0.45, 0.54) for $x = 0.030$, (0.43, 0.56) for $x = 0.035$, and (0.41, 0.58) for $x = 0.040$ (Figure 6b). All these emission colors fall into the red to green region of the CIE chromaticity diagram, and the films exhibit bright colors changing from red to orange, yellow, and then green as shown in Figure 6c. $Tb^{3+} \rightarrow Eu^{3+}$ energy transfer (ET) is well known in the phosphors codoped with Tb^{3+}/Eu^{3+} because of the substantial spectral overlap between the $^5D_4 \rightarrow {}^7F_J$ emissions of Tb^{3+} and the $^7F_{0,1} \rightarrow {}^5D_{0-2}$ excitation absorptions of Eu^{3+}, and the efficiency of ET can be analyzed when the Tb^{3+} content is fixed [31]. Similar analysis, however, can hardly be made in this work since both the Eu^{3+} and Tb^{3+} contents are variables.

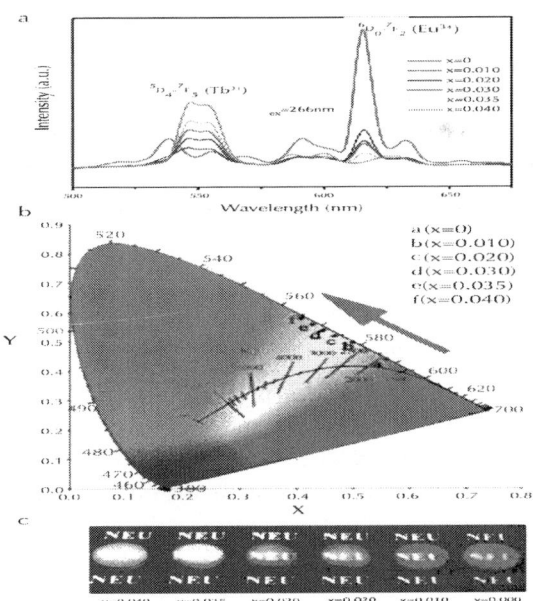

Figure 6: Photoluminescence spectra, CIE chromaticity diagram, and multi-color emission of the oxide films. Photoluminescence spectra (a) and CIE

chromaticity diagram (b) of the $(Y_{0.96}Tb_xEu_{0.04-x})_2O_3$ $(0 \leq x \leq 0.04)$ films. Part (c) shows multi-color emission of the oxide films under 254-nm irradiation from a hand-held UV lamp.

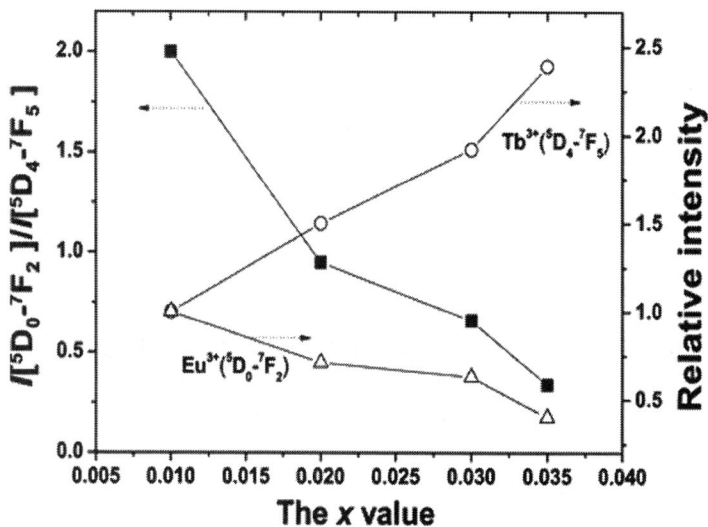

Figure 7: The relative intensities of two typical emission peaks. Correlation of the relative intensities $I(^5D_0 \rightarrow {}^7F_2)$ and $I(^5D_4 \rightarrow {}^7F_5)$ and the $I(^5D_0 \rightarrow {}^7F_2)/I(^5D_4 \rightarrow {}^7F_5)$ intensity ratio with Tb content for $(Y_{0.96}Tb_xEu_{0.04-x})_2O_3$. The relative intensities are obtained by normalizing the observed PL intensities to that of the $x = 0.01$ sample.

CONCLUSIONS

In this work, tens of micron-sized unilamellar nanosheets were efficiently exfoliated from well-crystallized $(Y_{0.96}Tb_xEu_{0.04-x})_2(OH)_5NO_3 \cdot nH_2O$ $(0 \leq x \leq 0.04)$ layered rare-earth hydroxides (LRHs) via hydrothermal anion exchange of the interlayer NO_3^- with much larger oleate anions, which were then employed for the construction of oriented fluorescent films. Detailed characterizations of the products by the combined techniques of FE-SEM, TEM, XRD, FT-IR, AFM, and PLE/PL have yielded the following main conclusions:

- Inserting water insoluble oleate anions ($C_{17}H_{33}COO^-$) into the interlayer of LRHs can be successfully achieved via hydrothermal processing (120°C for 24 h), which disorders the stacking of the *ab* plane along the *c* direction and weakens the interaction between the adjacent layers but with little damage to the *ab* plane. The intercalation of oleate extremely expands the interlayer distance up to approximately 5.2 nm, resulting in thickness increase of the LRH crystals from approximately 1 to 10 μm.

- Delamination of the oleate-inserted LRHs into unilamellar nanosheets with lateral sizes of ≥20 μm and a thickness of approximately 1.55 nm has been achieved by dispersing LRH-oleate in toluene, followed by slow stirring.

- Highly [111]-oriented and transparent films of $(Y_{0.96}Tb_xEu_{0.04-x})_2O_3$ $(0 \leq x \leq 0.04)$, with thicknesses of approximately 100 nm, have been constructed through spin-coating the colloidal nanosheets on quartz substrates, followed by calcination at 800°C. Upon UV excitation at 266 nm, the oxide films exhibit bright emissions and emission color can be tuned from red, orange, yellow, and then to green by increasing the Tb^{3+} content.

AUTHORS' CONTRIBUTIONS

QZ and JGL conceived the project and drafted the manuscript. QZ and ZXX carried out the experiments. XDL participated in the film preparation. YQ and XDS were involved in sample characterization and results discussion. All the authors have read and approved the final manuscript.

ACKNOWLEDGEMENTS

This work was supported in part by the National Natural Science Foundation of China (grants 51302032 and 51172038), the Fundamental Research Funds for the Central Universities (grants N140204002 and N110802001), the Liaoning Province Doctor Startup Fund (grant 20131035), and the Grant-in-Aid for Scientific Research (KAKENHI, no. 26420686). Thanks are due to the Materials Analysis Station of the National Institute for Materials Science (NIMS) for allowing access to research facilities.

REFERENCES

1. Ogawa M, Kuroda K: Photofunctions of intercalation compounds. *Chem Rev* 1995, 95:399-438.

2. Ida S, Ogata C, Eguchi M, Youngblood WJ, Mallouk TE, Matsumoto Y: Photoluminescence of perovskite nanosheets prepared by exfoliation of layered oxides, $K_2Ln_2Ti_3O_{10}$, $KLnNb_2O_7$, and $RbLnTa_2O_7$ (Ln: lanthanide ion).*J Am Chem Soc* 2008, 130:7052-9.

3. Ida S, Shiga D, Koinuma M, Matsumoto Y: Synthesis of hexagonal nickel hydroxide nanosheets by exfoliation of layered nickel hydroxide intercalated with dodecyl sulfate ions.*J Am Chem Soc* 2008, 130:14038-9.

4. Ida S, Ogata C, Shiga D, Izawa K, Ikeue K, Matsumoto Y: Dynamic control of photoluminescence for self-assembled nanosheet films intercalated with lanthanide ions by using a photoelectrochemical reaction.*Angew Chem Int Ed* 2008, 47:2480-3.

5. Ida S, Ogata C, Matsumoto Y: pH dependence of the photoluminescence of Eu^{3+}-intercalated layered titanium oxide.*J Phys Chem C* 2009, 113:1896-900.

6. Oh E-J, Kim TW, Lee KM, Song MS, Jee AY, Lim ST, *et al.*: Unilamellar nanosheet of layered manganese cobalt nickel oxide and its heterolayered film with polycations.*ACS Nano* 2010, 8:4437-44.

7. Liu ZP, Ma R, Osada M, Iyi N, Ebina Y, Takada K, *et al.*: Synthesis, anion exchange, and delamination of Co-Al layered double hydroxide: assembly of the exfoliated nanosheet/polyanion composite films and magneto-optical studies.*J Am Chem Soc* 2006, 128:4872-80.

8. Ma RZ, Sasaki T: Nanosheets of oxides and hydroxides: ultimate 2D charge-bearing functional crystallites.*Adv Mater* 2010, 22:5082-104.

9. Liu N, Luo F, Wu HX, Liu YH, Zhang C, Chen J: One-step ionic-liquid-assisted electrochemical synthesis of ionic-liquid-functionalized graphene sheets directly from graphite.*Adv Funct Mater* 2008, 18:1518-25.

10. Omomo Y, Sasaki T, Wang LZ, Watanabe M: Redoxable nanosheet crystallites of MnO_2 derived via delamination of a layered manganese oxide.*J Am Chem Soc* 2003, 125:3568-75.

11. Tanaka H, Okumiya T, Ueda SK, Taketani Y, Murakami M: Preparation of nanosheet by exfoliation of layered iron phenyl phosphate under ultrasonic irradiation.*Mater Res Bull* 2009, 44:328-33.

12. Izawa K, Ida S, Unal U, Yamaguchi T, Kang JH, Choy JH, *et al.*: A new approach for the synthesis of layered niobium sulfide and restacking route of NbS_2 nanosheet.*J Solid State Chem* 2008, 181:319-24.

13. Gándara F, Perles J, Snejko N, Iglesias M, Gómez-Lor B, Gutiérrez-Puebla E, *et al.*: Layered rare-earth hydroxides: a class of pillary crystalline compounds for intercalation chemistry.*Angew Chem Int Ed* 2006, 45:7998-8001.

14. McIntyre LJ, Jackson LK, Fogg AM: $Ln_2(OH)_5NO_3 \cdot xH_2O$ (Ln = Y, Gd-Lu): a novel family of anion exchange intercalation hosts. *Chem Mater* 2008, 20:335-40.

15. Geng FX, Ma RZ, Sasaki T: Anion-exchangeable layered materials based on rare-earth phosphors: unique comination of rare-earth host and exchangeable anions.*Acc Chem Res* 2010, 43:1177-85.

16. McIntyre LJ, Prior TJ, Fogg AM: Observation and isolation of layered and framework ytterbium hydroxide phases using in situ energy-dispersive X-ray diffraction.*Chem Mater* 2010, 22:2635-45.

17. Poudret L, Prior TJ, McIntyre LJ, Fogg AM: Synthesis and crystal structures of new lanthanide hydroxyhalide anion exchange materials, $Ln_2(OH)_5X \cdot 1.5H_2O$ (X = Cl, Br; Ln = Y, Dy, Er, Yb). *Chem Mater* 2008, 20:7447-53.

18. Lee KH, Byeon SH: Extended members of the layered rare-earth hydroxides family, $RE_2(OH)_5NO_3 \cdot nH_2O$ (RE = Sm, Eu, and Gd): synthesis and anion-exchange behavior.*Eur J Inorg Chem* 2009, 2009:929-36.

19. Lee KH, Byeon SH: Synthesis and aqueous colloidal solutions of $RE_2(OH)_5NO_3 \cdot nH_2O$ (RE = Nd and La).*Eur J Inorg Chem* 2009, 2009:4727-32.

20. Lee BI, Bae JS, Lee ES, Byeon SH: Synthesis and photoluminescence of colloidal solution containing layered rare-earth hydroxide nanosheets. *Bull Korean Chem Soc* 2012, 33:601-7.

21. Zhu Q, Li JG, Zhi CY, Li XD, Sun XD, Sakka Y, et al.: Layered rare-earth hydroxides (LRHs) of $(Y_{1-x}Eu_x)_2(OH)_5NO_3 \cdot nH_2O$ $(x = 0–1)$: structural variations by Eu^{3+} doping, phase conversion to oxides, and the correlation of photoluminescence behaviors. *Chem Mater* 2010, 22:4204-13.

22. Hu LF, Ma R, Ozawa TC, Sasaki T: Exfoliation of layered europium hydroxide into unilamellar nanosheets. *Chem Asian J* 2010, 5:248-51.

23. Lee KH, Lee BI, You JH, Byeon SH: Transparent Gd_2O_3:Eu phosphor layer derived exfoliated layered gadolinium hydroxide nanosheets. *Chem Commun* 2010, 46:1461-3.

24. Zhu Q, Li JG, Zhi CY, Ma R, Sasaki T, Xu JX, et al.: Nanometer-thin layered hydroxide platelets of $(Y_{0.95}Eu_{0.05})_2(OH)_5NO_3 \cdot xH_2O$: exfoliation-free synthesis, self-assembly, and the derivation of dense oriented oxide films of high transparency and greatly enhanced luminescence. *J Mater Chem* 2011, 21:6903-8.

25. Zhu Q, Li JG, Ma R, Sasaki T, Yang X, Li XD, et al.: Well-defined crystallites autoclaved from the nitrate/NH_4OH reaction system as the precursor for $(Y, Eu)_2O_3$ red phosphor: crystallization mechanism, phase and morphology control, and luminescent property. *J Solid State Chem* 2012, 192:229-37.

26. Zhu Q, Li JG, Li XD, Sun XD, Qi Y, Zhu MY, et al.: Tens of micron sized unilamellar nanosheets of Y/Eu layered rare-earth hydroxide (LRH): efficient exfoliation via fast anion exchange and their self-assembly into oriented oxide film with enhanced photoluminescence. *Sci Technol Adv Mater* 2014, 15:014203.

27. Nakamoto K: *Infrared spectra of inorganic and coordination compounds.* John Wiley & Sons, New York; 1963.

28. Gadsden JA: *Infrared spectra of minerals and related inorganic compounds.* Butterworth, Newton, MA; 1975.

29. Zhu Q, Li JG, Li XD, Sun XD, Sakka Y: Monodispersed colloidal spheres for $(Y, Eu)_2O_3$ red phosphors: establishment of processing window and size-dependent luminescence behavior. *Sci Technol Adv Mater* 2011, 12:055001.

30. Zhu Q, Li JG, Li XD, Sun XD: Morphology-dependent crystallization and luminescence behavior of $(Y, Eu)_2O_3$ red phosphors.*Acta Mater* 2009, 57:5975-85.

31. Wu XL, Li JG, Li JK, Zhu Q, Li XD, Sun XD, *et al.*: Layered rare-earth hydroxide (LRH) and oxide nanoplates of the Y/Tb/Eu system: phase controlled processing, structure characterization, and color-tunable photoluminescence via selective excitation and efficient energy transfer.*Sci Technol Adv Mater* 2013, 14:015006.

Characterization of Magnetic Nanoparticle by Dynamic Light Scattering

JitKang Lim[1, 2], Swee Pin Yeap[1], Hui Xin Che[1], and
Siew Chun Low[1]

[1]School of Chemical Engineering, Universiti Sains Malaysia, Nibong
Tebal, Penang, 14300, Malaysia
[2]Department of Physics, Carnegie Mellon University, Pittsburgh, PA,
15213, USA

ABSTRACT

Here we provide a complete review on the use of dynamic light
scattering (DLS) to study the size distribution and colloidal stability of

magnetic nanoparticles (MNPs). The mathematical analysis involved in obtaining size information from the correlation function and the calculation of Z-average are introduced. Contributions from various variables, such as surface coating, size differences, and concentration of particles, are elaborated within the context of measurement data. Comparison with other sizing techniques, such as transmission electron microscopy and dark-field microscopy, revealed both the advantages and disadvantages of DLS in measuring the size of magnetic nanoparticles. The self-assembly process of MNP with anisotropic structure can also be monitored effectively by DLS.

REVIEW

Introduction

Magnetic nanoparticles (MNPs) with a diameter between 1 to 100 nm have found uses in many applications [1,2]. This nanoscale magnetic material has several advantages that provide many exciting opportunities or even a solution to various biomedically [3-5] and environmentally [6-8] related problems. Firstly, it is possible to synthesize a wide range of MNPs with well-defined structures and size which can be easily matched with the interest of targeted applications. Secondly, the MNP itself can be manipulated by an externally applied magnetic force. The capability to control the spatial evolution of MNPs within a confined space provides great benefits for the development of sensing and diagnostic system/techniques [9,10]. Moreover MNPs, such as Fe^0 and Fe_3O_4, that exhibit a strong catalytic function can be employed as an effective nanoagent to remove a number of persistent pollutants from water resources [11,12]. In addition to all the aforementioned advantages, the recent development of various techniques and procedures for producing highly monodispersed and size-controllable MNPs [13,14] has played a pivotal role in promoting the active explorations and research of MNPs.

In all of the applications involving the use of MNPs, the particle size remained as the most important parameter as many of the chemical and physical properties associated to MNPs are strongly dependent upon the nanoparticle diameter. In particular, one of the unique features of a

MNP is its high-surface-to-volume ratio, and this property is inversely proportional to the diameter of the MNP. The smaller the MNP is, the larger its surface area and, hence, the more loading sites are available for applications such as drug delivery and heavy metal removal. Furthermore, nanoparticle size also determines the magnetophoretic forces (F_{mag}) experienced by a MNP since F_{mag} is directly proportional to the volume of the particles [15]. In this regard, having size information is crucial as at nanoregime, the MNP is extremely susceptible to Stoke's drag [16] and thermal randomization energy [17]. The successful manipulation of MNP can only be achieved if the F_{mag} introduced is sufficient to overcome both thermal and viscous hindrances [18]. In addition, evidences on the (eco)toxicological impacts of nanomaterials have recently surfaced[19]. The contributing factors of nanotoxicity are still a subject of debate; however, it is very likely due to either (1) the characteristic small dimensional effects of nanomaterials that are not shared by their bulk counterparts with the same chemical composition [20] or (2) biophysicochemical interactions at the nano-bio interface dictated by colloidal forces [21]. For either reason, the MNP's size is one of the determining factors.

The technique of dynamic light scattering (DLS) has been widely employed for sizing MNPs in liquid phase [22,23]. However, the precision of the determined particle size is not completely understood due to a number of unevaluated effects, such as concentration of particle suspension, scattering angle, and shape anisotropy of nanoparticles [24]. In this review, the underlying working principle of DLS is first provided to familiarize the readers with the mathematical analysis involved for correct interpretation of DLS data. Later, the contribution from various factors, such as suspension concentration, particle shape, colloidal stability, and surface coating of MNPs, in dictating the sizing of MNPs by DLS is discussed in detail. It is the intention of this review to summarize some of the important considerations in using DLS as an analytical tool for the characterization of MNPs.

Overview of Sizing Techniques for MNPs

There are numerous analytical techniques, such as DLS [25], transmission electron miscroscopy (TEM) [26], thermomagnetic measurement [27], dark-field microscopy [17,18], atomic force microscopy (AFM) [28], and acoustic spectrometry measurement [29], that have been employed to measure the size/size distribution of MNPs (Table 1). TEM is one of the most powerful analytical tools available which can give direct structural and size information of the MNP. Through the use of the short wavelengths achievable with highly accelerated electrons, it is capable to investigate the structure of a MNP down to the atomic level of detail, whereas by performing image analysis on the TEM micrograph obtained, it is possible to give quantitative results on the size distribution of the MNP. This technique, however, suffered from the small sampling size involved. A typical MNP suspension composed of 10^{10} to 10^{15} particles/mL and the size analysis by measuring thousands or even tens of thousands of particles still give a relatively small sample pool to draw statistically conclusive remarks.

Table 1: Common analytical techniques and the associated range scale involved for nanoparticle sizing

Techniques	Approximated working size range
Dynamic light scattering	1 nm to approximately 5 µm
Transmission electron microscopy	0.5 nm to approximately 1 µm
Atomic force microscopy	1 nm to approximately 1 µm
Dark-field microscopy	5 to 200 nm
Thermomagnetic measurement	10 to approximately 50 nm

Lim et al.

Lim et al. Nanoscale Research Letters 2013 8:381 doi:10.1186/1556-276X-8-381

Thermomagnetic measurement extracts the size distribution of an ensemble of superparamagnetic nanoparticles from zero-field cooling (ZFC) magnetic moment, $m_{ZFC}(T)$, data based on the Néel model [27]. This method is an indirect measurement of particle size and relies on the underlying assumption of the mathematical model used to calculate the size distribution. In addition, another limitation of this analytical method includes the magnetic field applied for ZFC measurements which must be small compared to the anisotropy field of the MNPs [30], and it also neglects particle-particle dipolar interactions which increase the apparent blocking temperature [31]. This technique, however, could give a very reliable magnetic size of the nanoparticle analyzed.

Dark-field microscopy relies on direct visual inspection of the optical signal emitted from the MNP while it undergoes Brownian motion. After the trajectories of each MNP over time t are recorded, the two-dimensional mean-squared displacement $<r^2> = 4Dt$ is used to calculate the diffusion coefficient D for each particle. Later on, the hydrodynamic diameters can be estimated via the Stokes-Einstein equation for the diffusion coefficients calculated for individual particles, averaging over multiple time steps [18]. Successful implementation of this technique depends on the ability to trace the particle optically by coating the MNP with a noble metal that exhibits surface Plasmon resonance within a visible wavelength. This extra synthesis step has significantly restricted the use of this technique as a standard route for sizing MNPs. The size of an MNP obtained through dark-field microscopy is normally larger than the TEM and DLS results [17]. It should be noted that dark-field microscopy can also be employed for direct visualization of a particle flocculation event [32]. As for AFM, besides the usual topographic analysis, magnetic imaging of a submicron-sized MNP grown on GaAs substrate has been performed with magnetic force microscopy equipment [33]. Despite all the recent breakthroughs, sample preparation and artifact observation are still the limiting aspect for the wider use of this technology for sizing MNPs [34].

The particle size and size distribution can also be measured with an acoustic spectrometer which utilizes the sound pulses transmitted through a particle suspension to extract the size-related information [29]. Based on the combined effect of absorption and scattering of acoustic energy, an acoustic sensor measures attenuation frequency spectra in the sample. This attenuation spectrum is used to calculate the particle size distribution. This technique has advantages over the light scattering method in studying samples with high polydispersity as the raw data for calculating particle size depend on only the third power of the particle size. This scenario makes contribution of the small (nano) and larger particles more even and the method potentially more sensitive to the nanoparticle content even in the very broad size distributions [35].

DLS, also known as photon correlation spectroscopy, is one of the most popular methods used to determine the size of MNPs. During the DLS measurement, the MNP suspension is exposed to a light beam (electromagnetic wave), and as the incident light impinges on the

MNP, the direction and intensity of the light beam are both altered due to a process known as scattering [36]. Since the MNPs are in constant random motion due to their kinetic energy, the variation of the intensity with time, therefore, contains information on that random motion and can be used to measure the diffusion coefficient of the particles [37]. Depending on the shape of the MNP, for spherical particles, the hydrodynamic radius of the particle R_H can be calculated from its diffusion coefficient by the Stokes-Einstein equation $D_f = k_B T/6\pi\eta R_H$, where k_B is the Boltzmann constant, T is the temperature of the suspension, and η is the viscosity of the surrounding media. Image analysis on the TEM micrographs gives the 'true radius' of the particles (though determined on a statistically small sample), and DLS provides the hydrodynamic radius on an ensemble average [38]. The hydrodynamic radius is the radius of a sphere that has the same diffusion coefficient within the same viscous environment of the particles being measured. It is directly related to the diffusive motion of the particles.

DLS has several advantages for sizing MNPs and has been widely used to determine the hydrodynamic size of various MNPs as shown in Table 2. First of all, the measuring time for DLS is short, and it is almost all automated, so the entire process is less labor intensive and an extensive experience is not required for routine measurement. Furthermore, this technique is non-invasive, and the sample can be employed for other purposes after the measurement. This feature is especially important for the recycle use of MNP with an expensive surface functional group, such as an enzyme or molecular ligands. In addition, since the scattering intensity is directly proportional to the sixth power of the particle radius, this technique is extremely sensitive towards the presence of small aggregates. Hence, erroneous measurement can be prevented quite effectively even with the occurrences of limited aggregation events. This unique feature makes DLS one of the very powerful techniques in monitoring the colloidal stability of MNP suspension.

Table 2: Hydrodynamic diameter of different MNPs determined by DLS

Type of MNPs	Surface coating	Hydrodynamic diameter by DLS (nm)	Reference

Fe0	Carboxymethyl cellulose	15-19	[39]
	Guar gum	350-700	[40]
	Poly(methacrylic acid)-poly(methyl methacrylate)-poly(styrenesulfonate) triblock copolymer	100-600	[41]
	Poly(styrene sulfonate)	30-90	[22]
-Fe2O3	Oleylamine or oleic acid	5-20	[42]
	Poly(N,N-dimethylacrylamide)	55-614	[43]
	Poly(ethylene oxide)-block-poly(glutamic acid)	42-68	[44]
	Poly(ethylene imine)	20-75	[45]
	Poly(-caprolactone)	193 ± 7	[46]
Fe3O4	Phospholipid-PEG	14.7 ± 1.4	[47]
	Polydimethylsiloxane	41.2 ± 0.4	[48]
	Oleic acid-pluronic	50-600	[49]
	Polyethylenimine (PEI)	50-150	[23,50]
	Polythylene glycol	10-100	[51]
	Triethylene glycol	16.5 ± 3.5	[52]
	Poly(N-isopropylacrylamide)	15-60	[53]
	Pluronic F127	36	[54]
	Poly(sodium 4-styrene sulfonate)	~200	[55]
	Poly(diallyldimethylammonium chloride)	107.4 ± 53.7	[56]
FePt	Poly(diallyldimethylammonium chloride)	30-100	[57]
NiO	Cetyltrimethyl ammonium bromide	10-80	[58]
	Fetal bovine serum	39.05	[59]
	Not specified	750 ± 30	[60]
CoO, Co2O3	Poly(methyl methacrylate)	59-85	[61]
CoFe	Hydroxamic and phosphonic acids	6.5-458.7	[62]

Lim et al.

Lim et al. Nanoscale Research Letters 2013 8:381 doi:10.1186/1556-276X-8-381

The Underlying Principle of DLS

The interaction of very small particles with light defined the most fundamental observations such as why is the sky blue. From a technological perspective, this interaction also formed the underlying working principle of DLS. It is the purpose of this section to describe the mathematical analysis involved to extract size-related information from light scattering experiments.

The Correlation Function

DLS measures the scattered intensity over a range of scattering angles θ_{dls} for a given time t_k in time steps Δt. The time-dependent intensity $I(q, t)$ fluctuates around the average intensity $I(q)$ due to the Brownian motion of the particles [38]:

$$[I(q)] = \lim_{t_k \to \infty} 1/t_k \int_0^{t_k} I(q,t) \cdot dt \approx \lim_{k \to \infty} \frac{1}{k} \Sigma_{i=1}^k I(q, i \cdot \Delta t)$$

(1)

where $[I(q)]$ represents the time average of $I(q)$. Here, it is assumed that tk, the total duration of the time step measurements, is sufficiently large such that $I(q)$ represents average of the MNP system. In a scattering experiment, normally, θ_{dls} (see Figure 1) is expressed as the magnitude of the scattering wave vector q as

$$q = (4\pi n/\lambda) \sin(\theta_{dls}/2)$$

(2)

where n is the refractive index of the solution and λ is the wavelength in vacuum of the incident light. Figure 2a illustrates typical intensity fluctuation arising from a dispersion of large particles and a dispersion of small particles. As the small particles are more susceptible to random forces, the small particles cause the intensity to fluctuate more rapidly than the large ones.

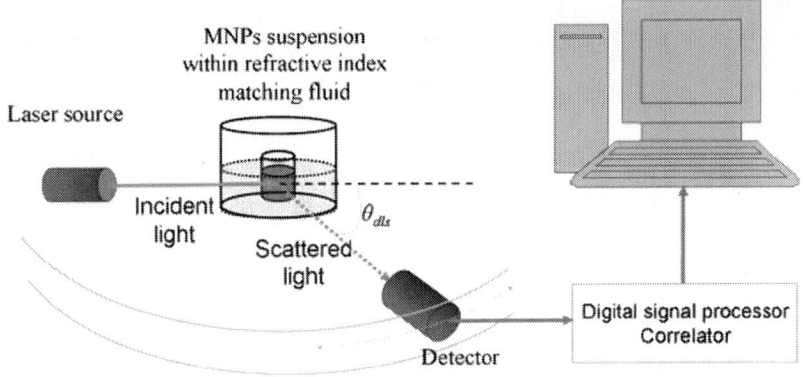

Figure 1: Optical configuration of the typical experimental setup for dynamic light scattering measurements. The setup can be operated at multiple angles.

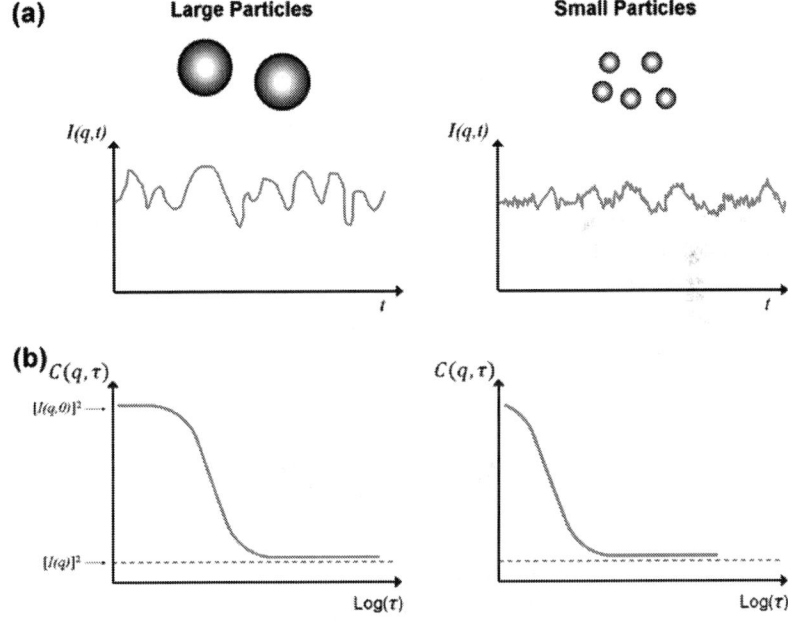

Figure 2: Schematic illustration of intensity measurement and the corresponding autocorrelation function in dynamic light scattering. The figure illustrates dispersion composed of large and small particles.(a) Intensity fluctuation of scattered light with time, and (b) the variation of autocorrelation function with delay time.

The time-dependent intensity fluctuation of the scattered light at a particular angle can then be characterized with the introduction of the autocorrelation function as

$$c(q,\tau) = \lim_{t_k \to \infty} 1/t_k \int_0^{t_k} I(q,t) \cdot I(q,t+\tau) \cdot dt$$

$$\approx \lim_{k \to \infty} \frac{1}{k} \Sigma_{j=0}^k I(q,i \cdot \Delta t) \cdot I(q,(i+j) \cdot \Delta t)$$

(3)

where $\tau = i\,\Delta t$ is the delay time, which represents the time delay between two signals $I(q,i\,\Delta t)$ and $I(q,(i+j)\,\Delta t)$. The function $C(q,\tau)$ is obtained for a series of τ and represents the correlation between the intensity at t_1 ($I(q,t_1)$) and the intensity after a time delay of τ ($I(q,t_1+\tau)$). The last part of the equation shows how the autocorrelation function is calculated experimentally when the intensity is measured in discrete time steps [37]. As for nanoparticle dispersion, the autocorrelation function decays more rapidly for small particles than for the large particles as depicted in Figure 2b. The autocorrelation function has its highest value of $[I(q,0)]^2$ at $\tau = 0$. As τ becomes sufficiently large at long time scales, the fluctuations becomes uncorrelated and $C(q,\tau)$ decreases to $[I(q)]^2$. For non-periodic $I(q,t)$, a monotonic decay of $C(q,\tau)$ is observed as τ increases from zero to infinity and

$$C(q,\tau)/{[I(q)]^2} = g^{(2)}(q,\tau) = 1 + \xi \left| g^{(1)}(q,\tau) \right|^2$$

(4)

where ξ is an instrument constant approximately equal to unity and $g^{(1)}(q,\tau)$ is the normalized electric field correlation function [63]. Equation 4 is known as the Siegert relation and is valid except in the case of scattering volume with a very small number of scatterers or when the motion of the scatterers is limited. For monodisperse, spherical particles, $g^{(1)}(\tau)$ is given by

$$g^{(1)}(q,\tau) = \exp\left(-D_f q^2 \tau\right).$$

$$(5)$$

Once the value of Df is obtained, the hydrodynamic diameter of a perfectly monodisperse dispersion composed of spherical particles can be inferred from the Stokes-Einstein equation. Practically, the correlation function observed is not a single exponential decay but can be expressed as

$$g^{(1)}(q,\tau) = \int_0^\infty G(\Gamma)e^{-\Gamma\tau}d\Gamma$$

$$(6)$$

where $G(\Gamma)$ is the distribution of decay rates Γ. For a narrowly distributed decay rate, cumulant method can be used to analyze the correlation function. A properly normalized correlation function can be expressed as

$$\ln\left(g^{(1)}(q,\tau)\right) = -\langle\Gamma\rangle\tau + \frac{\mu_2}{2}\tau^2$$

$$(7)$$

where $\langle\Gamma\rangle$ is the average decay rate and can be defined as

$$\langle\Gamma\rangle = \int_0^\infty G(\Gamma)\Gamma d\Gamma$$

$$(8)$$

and $\mu_2 = \langle\Gamma\rangle^2 - \langle\Gamma\rangle^2$ is the variance of the decay rate distribution. Then, the polydispersity index (PI) is defined as PI $= \mu_2/\langle\Gamma\rangle^2$. The average hydrodynamic radius is obtained from the average decay rate $\langle\Gamma\rangle$ using the relation

$$R_H = \frac{k_B T}{6\pi\eta\langle\Gamma\rangle}q^2$$

(9)

Z-average

In most cases, the DLS results are often expressed in terms of the Z-average. Since the Z-average arises when DLS data are analyzed through the use of the cumulant technique [64], it is also known as the "cumulant mean." Under Rayleigh scattering, the amount of light scattered by a single particle is proportional to the sixth power of its radius (volume squared). This scenario causes the averaged hydrodynamic radius determined by DLS to be also weighted by volume squared. Such an averaged property is called the Z-average. For particle suspension with discrete size distribution, the Z-average of some arbitrary property y would be calculated as

$$\langle y\rangle = \frac{\Sigma_i n_i R_{H,i}^6 y_i}{\Sigma_i n_i R_{H,i}^6}$$

(10)

where n_i is the number of particles of type i having a hydrodynamic radius of $R_{H,i}$ and property y. If we assume that this particle dispersion consists of exactly two sizes of particles 1 and 2, then Equation 10 yields

$$\langle y\rangle = \frac{n_1 R_{H,1}^6 y_1 + n_2 R_{H,2}^6 y_2}{n_1 R_{H,1}^6 + n_2 R_{H,2}^6}$$

(11)

where $R_{H,i}$ and y_i are the volume and arbitrary property for particle 1 ($i = 1$) and particle 2 ($i = 2$). Suppose that two particles 1 combined to form one particle 2 and assume that we start with n_0 total of particle 1, some of which combined to form n_2 number of particle 2. With this assumption, we have $n_1 = n_0 - n_2$ number of particle 1. Moreover, under this assumption $R_{H,2} = 2\, R_{H,1}$. Substitute these relations into Equation 11; then, the Z-average of property y becomes

$$\frac{\langle y \rangle}{y_1} = \frac{1 + \left(2\left(\frac{y_2}{y_1}\right) - 1\right)2\left(\frac{n_2}{n_0}\right)}{1 + 2\left(\frac{n_2}{n_0}\right)}$$

(12)

where $2n_2/n_0$ is the fraction of total particle 1 existing as particle 2. Solving this fraction, we obtained

$$\frac{2n_2}{n_0} = \frac{\frac{\langle y \rangle}{y_1} - 1}{\frac{2y_2}{y_1} - \frac{\langle y \rangle}{y_1} - 1}$$

(13)

However, it should be noted that Z-average should only be employed to provide the characteristic size of the particles if the suspension is monomodal (only one peak), spherical, and monodisperse. As shown in Figure 3, for a mixture of particles with obvious size difference (bimodal distribution), the calculated Z-average carries irrelevant size information.

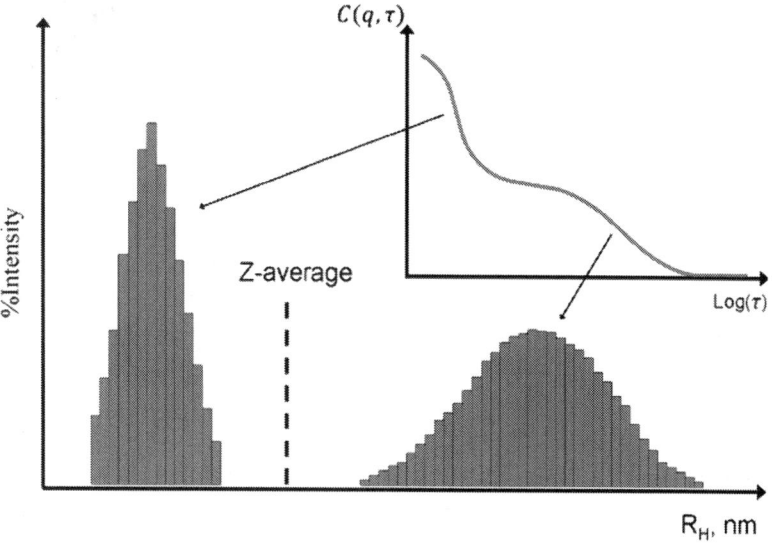

Figure 3: Z-average (cumulant) size for particle suspension with bimodal distribution.

DLS Measurement of MNPs

The underlying challenges of measuring the size of MNPs by DLS lay in the facts that (1) for engineering applications, these particles are typically coated with macromolecules to enhance their colloidal stability (see Figure 4) and (2) there present dipole-dipole magnetic interactions between the none superparamagnetic nanoparticles. Adsorbing macromolecules onto the surface of particles tends to increase the apparent R_H of particles. This increase in R_H is a convenient measure of the thickness of the adsorbed macromolecules [65]. This section is dedicated to the scrutiny of these two phenomena and also suspension concentration effect in dictating the DLS measurement of MNPs. All DLS measurements were performed with a Malvern Instrument Zetasizer Nano Series (Malvern Instruments, Westborough, MA, USA) equipped with a He-Ne laser (λ = 633 nm, max 5 mW) and operated at a scattering angle of 173°. In all measurements, 1 mL of particle suspensions was employed and placed in a 10 mm × 10 mm quartz cuvette. The iron oxide MNP used in this study was synthesized by a high-temperature decomposition method [17].

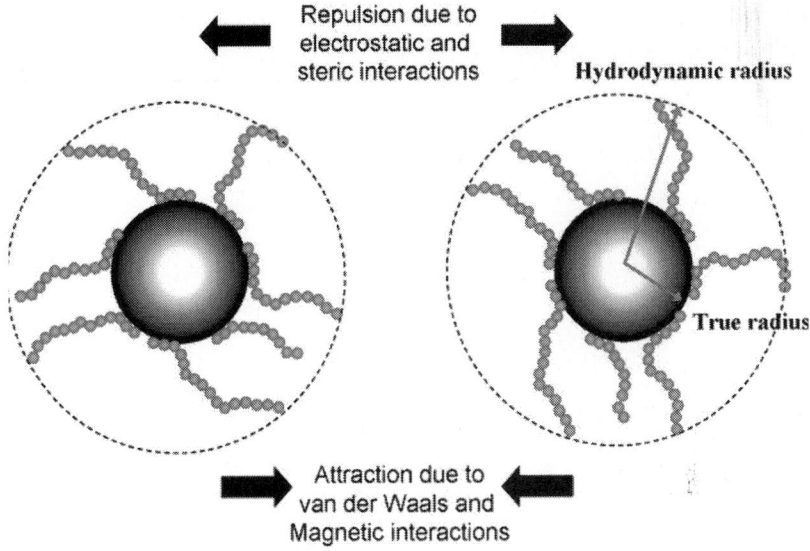

Figure 4: Pictorial representation of two MNPs and major interactions. The image shows two MNPs coated with macromolecules with repeated segments and the major interactions involved between them in dictating the colloidal stability of MNP suspension.

Size Dependency of MNP in DLS Measurement

In order to demonstrate the sizing capability of DLS, measurements were conducted on three species of Fe_3O_4 MNPs produced by high-temperature decomposition method which are surface modified with oleic acid/oleylamine in toluene (Figure 5). The TEM image analyses performed on micrographs shown in Figure 5 (from top to bottom) indicate that the diameter of each particle species is 7.2 ± 0.9 nm, 14.5 ± 1.8 nm, and 20.1 ± 4.3 nm, respectively. The diameters of these particles obtained from TEM and DLS are tabulated in Table 3. It is very likely that the main differences between the measured diameters from these two techniques are due to the presence of an adsorbing layer, which is composed of oleic acid (OA) and oleylamine (OY), on the surface of the particle. Small molecular size organic compounds, such as OA and OY, are electron transparent, and therefore, they did not show up in the TEM micrograph (Figure 5). Given that the chain lengths of OA and OY are approximately 2 nm [66,67], the best match of DLS and TEM, in terms of measured diameter, can be observed from middle-sized Fe_3O_4 MNPs.

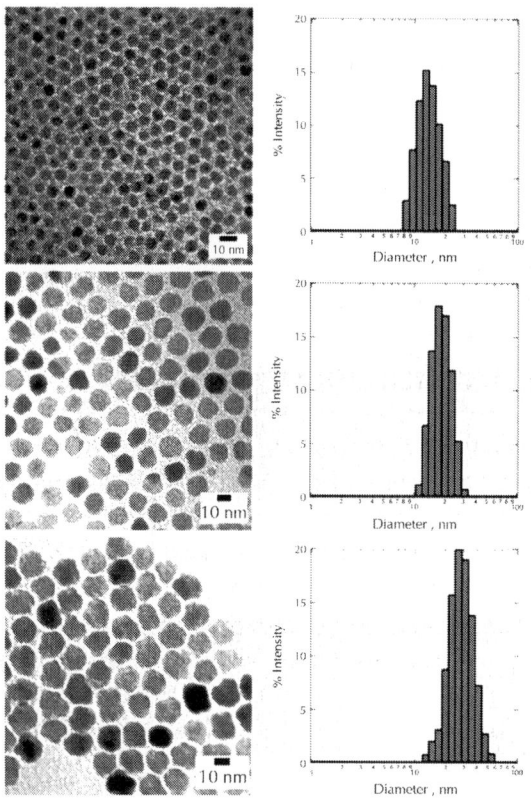

Figure 5: TEM micrographs of Fe3O4 MNPs with their size distribution determined by DLS. The Z-average of MNP calculated from the DLS data is (top) 16.9 ± 5.2 nm, (middle) 21.1 ± 5.5 nm, and (bottom) 43.1 ± 14.9 nm, respectively.

Table 3: Diameter of Fe3O4 MNP determined by TEM and DLS (Z -average)

Particle	TEM (nm)	DLS (nm)	Difference (nm)
Fe3O4	7.2	16.9	9.7
	14.5	21.1	6.6
	20.1	43.1	23.0

Lim et al.

Lim et al. Nanoscale Research Letters 2013 8:381 doi:10.1186/1556-276X-8-381

For small-sized MNPs, the radius of curvature effect is the main contributing factor for the large difference observed on the averaged diameter from DLS and TEM. This observation has at least suggested that for any inference of layer thickness from DLS measurement, the particles with a radius much larger than the layer thickness should be employed. In this measurement, the fractional error in the layer thickness can be much larger than the fractional error in the radius with the measurement standard deviation of only 0.9 nm for TEM but at a relatively high value of 5.2 nm for DLS. At a very large MNP size of around 20 nm (bottom image of Figure 5), the Z-average hydrodynamic diameter is 23 nm larger than the TEM size. Moreover, the standard deviation of the DLS measurement of this particle also increased significantly to 14.9 nm compared to 5.2 and 5.5 nm for small- and middle-sized MNPs, respectively. This trend of increment observed in standard deviation is consistent with TEM measurement. Both the shape irregularity and polydispersity, which are the intrinsic properties that can be found in a MNP with a diameter of 20 nm or above, contribute to this observation. For a particle larger than 100 nm, other factors such as electroviscous and surface roughness effects should be taken into consideration for the interpretation of DLS results [68].

MNP Concentration Effects

In DLS, the range of sample concentration for optimal measurements is highly dependent on the sample materials and their size. If the sample is too dilute, there may be not enough scattering events to make a proper measurement. On the other hand, if the sample is too concentrated, then multiple scattering can occur. Moreover, at high concentration, the particle might not be freely mobile with its spatial displacement driven solely by Brownian motion but with the strong influences of particle interactions. This scenario is especially true for the case of MNPs with interparticle magnetic dipole-dipole interactions.

Figure 6 illustrates the particle concentration effects on 6- and 18-nm superparamagnetic iron oxide MNPs, with no surface coating, dispersed in deionized water. Both species of MNPs show strong concentration dependency as their hydrodynamic diameter increases with the concentration increment. The hydrodynamic diameter for small particles increases from 7.1 ± 1.9 nm to 13.2 ± 3.3 nm as the MNP concentration increases from 25 to 50 mg/L. On the other hand, the hydrodynamic diameter of large particles remains to be quite constant until around 100 mg/L and then only experiences a rapid jump of the detected size from 29.3 ± 4.6 nm (at 100 mg/L) to 177.3 ± 15.8 nm (at 250 mg/L). Since the concentration of the MNP is prepared in mass basis, the presence of an absolute number of particles in a given volume of solution is almost two orders of magnitude higher in a small-particle suspension. For example, at 100 mg/L, the concentrations for small and larger particles are calculated as 1.7×10^{20} particles (pts)/m^3 and 6.3×10^{18} pts/m^3 by assuming that the composition material is magnetite with a density of 5.3 g/cm^3. This concentration translated to a collision frequency of 85,608 s^{-1} and 1,056 s^{-1}. So, at the same mass concentration, it is more likely for small particles to experience the non-self-diffusion motions.

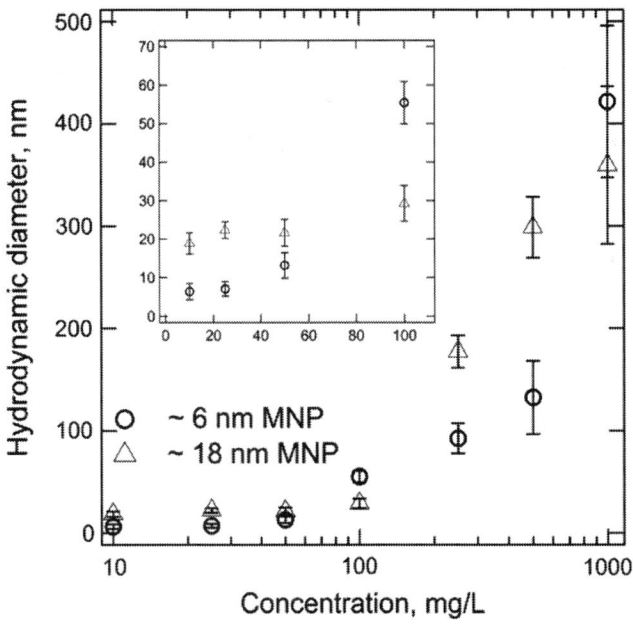

Figure 6: Particle concentration effects on the measurement of hydrodynamic diameter by DLS.

For both species of particles, the upward trends of hydrodynamic diameter, which associates to the decrement of diffusion coefficient, reflect the presence of a strong interaction between the particles as MNP concentration increases. Furthermore, since the aggregation rate has a second-order dependency on particle concentration [69], the sample with high MNP concentration has higher tendency to aggregate, leading to the formation of large particle clusters. Therefore, the initial efforts for MNP characterization by using DLS should focus on the determination of the optimal working concentration.

Colloidal Stability of MNPs

Another important use of DLS in the characterization of MNPs is for monitoring the colloidal stability of the particles [70]. An iron oxide MNP coated with a thin layer of gold with a total diameter of around 50 nm is further subjected for surface functionalization by a variety of macromolecules [65]. The colloidal stability of the MNP coated with all these macromolecules suspended in 154 mM ionic strength phosphate buffer solution (PBS) (physiologically relevant environment for biomedical application) is monitored by DLS over the course of 5 days (Figure 7). The uncoated MNP flocculated immediately after their introduction to PBS and is verified with the detection of micron-sized objects by DLS.

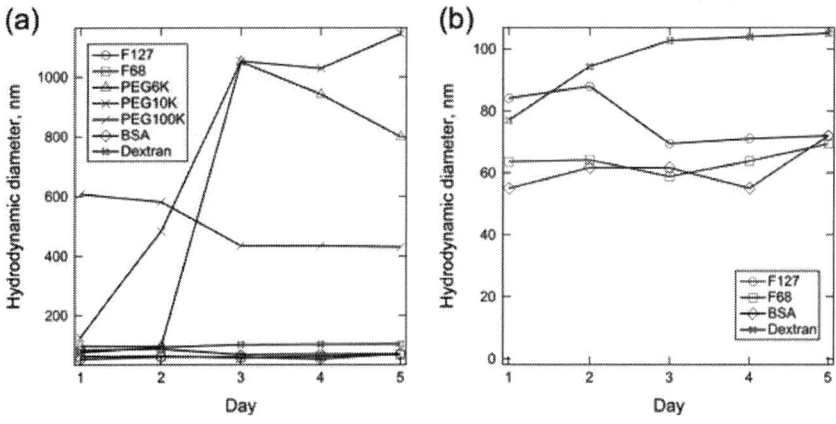

Figure 7: Intensity-weighted average hydrodynamic diameter for core-shell nanoparticles with different adsorbed macromolecules in PBS. (a) Extensive

aggregation is evident with PEG 6k, PEG10k, and PEG100k, while (b) bovine serum albumin (BSA), dextran, Pluronic F127, and Pluronic F68 provided stable hydrodynamic diameters over the course of 5 days. 'Day 0' corresponds to the start of the overnight adsorption of macromolecules to the MNPs. Copyright 2009 American Chemical Society. Reprinted with permission from [65].

As shown in Figure 7, both polyethylene glycol (PEG) 6k and PEG 10k are capable of tentatively stabilizing the MNPs in PBS for the first 24 and 48 h. Aggregation is observed with the detection of particle clusters with a diameter of more than 500 nm. After this period of relative stability, aggregation accelerated to produce micron-sized aggregates by day 3. Actually, the continuous monitoring of MNP size by DLS after this point is less meaningful as the dominating motion is the sedimentation of large aggregates [71]. For PEG 6k and PEG 10k that have a rather low degree of polymerization, the loss of stability over a day or two could have been due to slow PEG desorption that would not be expected of larger polymers. Nevertheless, PEG 100k-coated MNPs were not as well stabilized as the PEG 6k- or PEG 10k-coated ones, despite the higher degree of polymerization that one might expect to produce greater adsorbed layer thicknesses and therefore longer-ranged steric forces. In addition to the degree of polymerization, as discussed by Golas and coworkers[72], the colloidal stability of polymeric stabilized MNPs is also dependent on other structural differences of the polymer employed, such as the chain architecture and the identity of the charged functional unit. In their work, DLS was used to confirm the nanoparticle suspensions that displayed the least sedimentation which was indeed stable against aggregation.

In addition to the popular use of DLS in sizing individual MNPs, this analytical technique is also being employed to monitor the aggregation behavior of MNPs and the size of final clusters formed[55,73]. The study of particle aggregates is important since the magnetic collection is a cooperative phenomenon [74,75]. Subsequently, it is much easier to harvest submicron-sized MNP clusters than individual particles. Hence, a magnetic nanocluster with loss-packed structure and uniform size and shape has huge potential for various engineering applications in which the real-time separation is the key requirement [76]. Therefore, the use of DLS to monitor the aggregation kinetic of MNPs is important to provide direct feedback about the time scale associated with this process [55,77]. Figure 8 illustrates the aggregation behavior of three species of 40-nm reactive nanoscale iron particles (RNIP),

27.5-nm magnetite (Fe_3O_4) MNP, and 40-nm hematite (α-Fe_2O_3) MNP [73]. Phenrat and coworkers have demonstrated that DLS can be an effective tool to probe the aggregation behavior of MNPs (Figure 8a). The time evolution of the hydrodynamic radius of these particles from monomodal to bimodal distribution revealed the aggregation kinetic of the particles. Together with the in situ optical microscopy observation, the mechanism of aggregation is proposed as the transitions from rapidly moving individual MNPs to the formation of submicron clusters that lead to chain formation and gelation (Figure 8b). By the combination of small-angle neutron scattering and cryo-TEM measurements, DLS can also be used as an effective tool to understand the fractal structure of this aggregate [78].

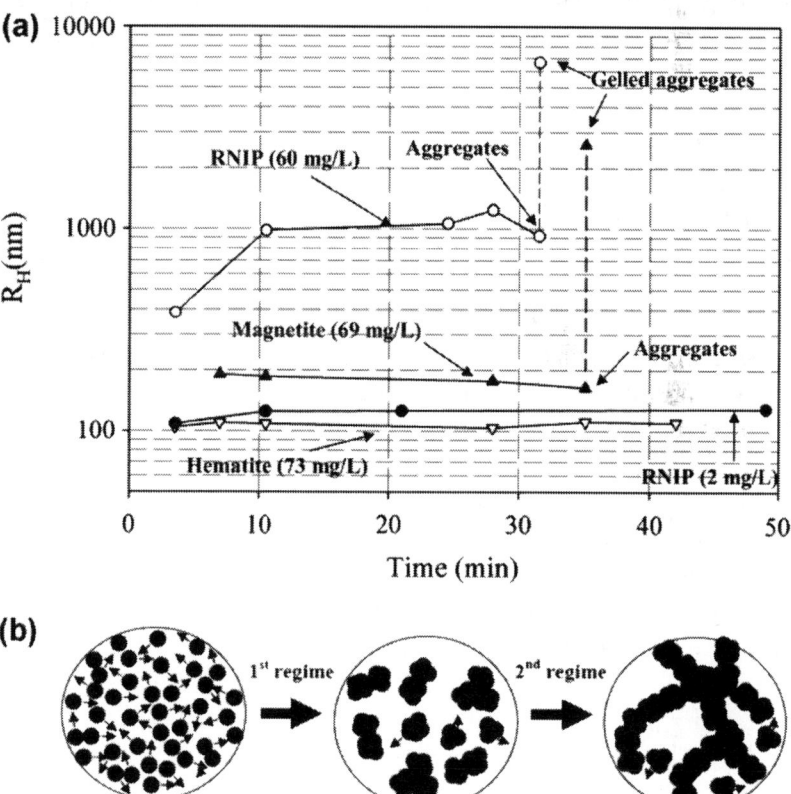

Figure 8: Evolution of hydrodynamic radius and MNP aggregation and gelation. (a) Evolution of the average hydrodynamic radius of dominant size class

of MNPs as a function of time for RNIP (Fe^0/Fe_3O_4), magnetite, and hematite at pH 7.4. The particle size distribution for RNIP and magnetite becomes bimodal at the last measured point due to gelation of aggregates. (b) Rapid MNP aggregation and subsequent chain-like gelation: rapid aggregation of MNP to form micron-sized clusters (first regime) and chain-like aggregation and gelation of the micron-sized aggregates (second regime). Copyright 2007 American Chemical Society. Reprinted with permission from [73].

DLS Measurement of Non-Spherical MNPs

Even though, under most circumstances, a more specialized analytical technique known as depolarized dynamic light scattering is needed to investigate the structural contribution of anisotropic materials [79], it is still possible to extract useful information for rod-like MNPs by conventional DLS measurement [80,81]. For rod-like particles, the decay rate in Equation 6 can be defined as

$$\Gamma = q^2 D_T + 6 D_R$$

(14)

where in a plot of Γ vs q^2, the value of rotational diffusion D_R can be obtained directly by an extrapolation of q to zero and the value of translational diffusion D_T from the slope of the curve[79]. For rigid non-interacting rods at infinite dilution with an aspect ratio (L/d) greater than 5, D_R and D_T can be expressed using Broersma's relations [82,83] or the stick hydrodynamic theory[84]. By performing angle-dependent DLS analysis on rod-like β-FeOOH nanorods as shown in Figure 9a, we found that the decay rate is linearly proportional to q^2 and passes through the origin (Figure 9b), suggesting that the nanorod motion is dominated by translational diffusion [85]. From Figure 9b, the slope of the graph yields the translational diffusion coefficient, $D_T = 7 \times 10^{-12}$ m²/s. This value of D_T corresponds to an equivalent spherical hydrodynamic diameter of 62.33 nm, suggesting that the DLS results with a single fixed angle of 173° overestimated the true diameter[86]. By taking the length and width of the nanorods as 119.7 and 17.5 nm (approximated from TEM images in Figure 9a), the D_T calculated by the stick hydrodynamic theory and Broersma's relationship is 7.09 ×

10^{-12} m²/s and 6.84×10^{-12} m²/s, respectively, consistent with the DLS results.

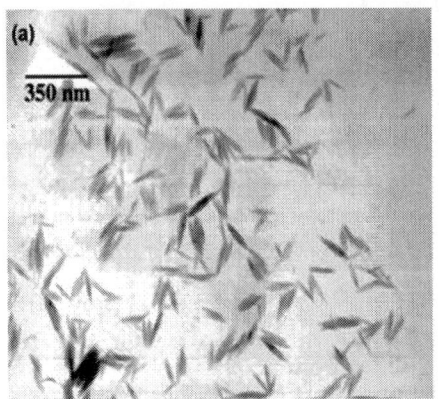 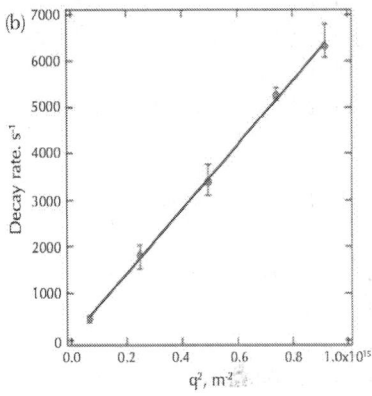

Figure 9: TEM images and graph of decay rate. (a) TEM images of β-FeOOH nanorods and (b) angle-dependent decay rate Γ of the nanorod showing a linear trend. Copyright 2009 Elsevier. Reprinted with permission from [86].

Since the β-FeOOH nanorods are self-assembled in a side-by-side fashion to form highly oriented 2-D nanorod arrays and the 2-D nanorod arrays are further stacked in a face-to-face fashion to form the final 3-D layered architectures, DLS can serve as an effective tool to monitor these transient behaviors [87]. Figure 10a depicts the structural changes of self-assembled nanorods over a time course of 7 h. To monitor the in situ real-time behavior of this self-assembly process, DLS was employed to provide the size distribution of the intermediate products that formed in the solution (Figure 10b). The temporal evolution of the detected size from 60 to 70 nm, to dual peaks, to eventually only a single distribution with a peak value of 700 nm indicating that all the building blocks are self-assembled into the large aggregates within the experiment time frame agrees well with the SEM observation (Figure 10a). This kinetic data time scale is involved in the full assembly of anisotropic nanomaterials from single building blocks to 2-D arrays and, eventually, 3-D micron-sized assemblies.

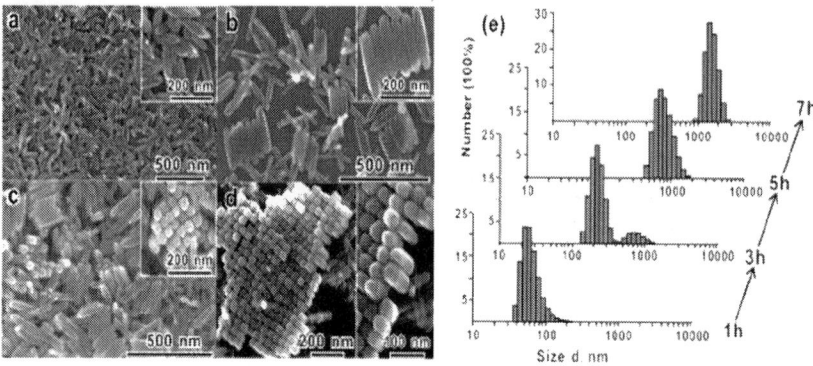

Figure 10: SEM images of the morphological evolution in the time-dependent experiments. (a) 1 h, (b) 3 h, (c) 5 h, and (d) 7 h. (e) Size distribution of the products obtained in the time-dependent experiments was monitored by DLS with the number averaged. Copyright 2010 American Chemical Society. Reprinted with permission from [87].

CONCLUSIONS

Dynamic light scattering is employed to monitor the hydrodynamic size and colloidal stability of the magnetic nanoparticles with either spherical or anisotropic structures. This analytical method cannot be employed solely to give feedbacks on the structural information; however, by combining with other electron microscopy techniques, DLS provides statistical representative data about the hydrodynamic size of nanomaterials. In situ, real-time monitoring of MNP suspension by DLS provides useful information regarding the kinetics of the aggregation process and, at the same time, gives quantitative measurement on the size of the particle clusters formed. In addition, DLS can be a powerful technique to probe the layer thickness of the macromolecules adsorbed onto the MNP. However, the interpretation of DLS data involves the interplay of a few parameters, such as the size, concentration, shape, polydispersity, and surface properties of the MNPs involved; hence, careful analysis is needed to extract the right information.

AUTHORS' CONTRIBUTIONS

JKL synthesized the MNPs, carried out TEM analysis, and drafted the manuscript. SPY carried out DLS measurement and data analysis. HXC carried out DLS measurement and data analysis. SCL participated in the design of the study and drafted the manuscript. All authors read and approved the final manuscript.

ACKNOWLEDGEMENTS

This material is based on the work supported by Research University (RU) (grant no. 1001/PJKIMIA/811219) from Universiti Sains Malaysia (USM), Exploratory Research Grants Scheme (ERGS) (grant no. 203/PJKIMIA/6730013) from the Ministry of Higher Education of Malaysia, and eScience Fund (grant no. 205/PJKIMIA/6013412) from MOSTI Malaysia. JKL and SWL are affiliated to the Membrane Science and Technology Cluster of USM.

REFERENCES

1. Lu AH, Salabas EL, Schüth F: Magnetic nanoparticles: synthesis, protection, functionalization, and application. *Angew Chem Int Ed* 2007, 46:1222-1244.

2. Pankhurst QA, Connolly J, Jones SK, Dobson J: Applications of magnetic nanoparticles in biomedicine. *J Phys D Appl Phys* 2003, 36:R167.

3. Adolphi NL, Huber DL, Bryant HC, Monson TC, Fegan DL, Lim JK, Trujillo JE, Tessier TE, Lovato DM, Butler KS, Provencio PP, Hathaway HJ, Majetich SA, Larson RS, Flynn ER:Characterization of single-core magnetite nanoparticles for magnetic imaging by SQUID relaxometry. *Phys Med Biol* 2010, 55:5985-6003.

4. Gupta AK, Gupta M: Synthesis and surface engineering of iron oxide nanoparticles for biomedical applications. *Biomaterials* 2005, 25:3995-4021.

5. Hao R, Xing R, Xu Z, Hou Y, Gao S, Sun S: Sythesis, functionalization and biomedical applications of multifunctional magnetic nanoparticles. *Adv Mater* 2010, 22:2729-2742.

6.   Cumbat L, Greenleaf J, Leun D, SenGupta AK: Polymer supported inorganic nanoparticles: characterization and environmental applications. *React Funct Polym* 2003, 54:167-180.

7.   Yantasee W, Warner CL, Sangvanich T, Addleman RS, Carter TG, Wiacek RJ, Fryxell GE, Timchalk C, Warner MG: Removal of heavy metals from aqueous systems with thiol functionalized superparamagnetic nanoparticles. *Environ Sci Technol* 2007, 41:5114-5119.

8.   Hu J, Lo IMC, Chen G: Comparative study of various magnetic nanoparticles for Cr(VI) removal. *Sep Purif Technol* 2007, 56:249-256.

9.   Dobson J: Remote control of cellular behavior with magnetic nanoparticles. *Nat Nanotech* 2008, 3:139-143.

10.   Gao J, Zhang W, Huang P, Zhang B, Zhang X, Xu B: Intracellular spatial control of fluorescent magnetic nanoparticles. *J Am Chem Soc* 2008, 130:3710-3711.

11.   Fiedor JN, Bostick WD, Jarabek RJ, Farrell J: Understanding the mechanism of uranium removal from groundwater by zero-valent iron using X-ray photoelectron spectroscopy. *Environ Sci Technol* 1998, 32:1466-1473.

12.   Feng J, Hu X, Yue PL, Zhu HY, Lu GQ: Degradation of azo-dye orange II by a photoassisted Fenton reaction using a novel composite of iron oxide and silicate nanoparticles as a catalyst. *Ind Eng Chem Res* 2003, 42:2058-2066.

13.   Sun S: Recent advances in chemical synthesis, self-assembly, and applications of FePt nanoparticles. *Adv Mater* 2006, 18:393-403.

14.   Park J, Joo J, Kwon SG, Jang Y, Hyeon T: Synthesis of monodisperse spherical nanocrystals. *Angew Chem Int Ed* 2007, 46:4630-4660.

15.   Zborowski M, Sun L, Moore LR, Williams PS, Chalmers JJ: Continuous cell separation using novel magnetic quadrupole flow sorter. *J Magn Magn Mater* 1999, 194:224-230.

16.   Purcell EM: Life at low Reynolds number. *Am J Phys* 1977, 45:3-11.

17.   Lim JK, Eggeman A, Lanni F, Tilton RD, Majetich SA: Synthesis and single-particle optical detection of low-polydispersity plasmonic-superparamagnetic nanoparticles. *Adv Mater* 2008, 20:1721-1726.

18. Lim JK, Lanni C, Evarts ER, Lanni F, Tilton RD, Majetich SA: Magnetophoresis of nanoparticles. *ACS Nano* 2011, 5:217-226.

19. Nel A, Xia T, Mädler L, Li N: Toxic potential of materials at the nanolevel. *Science* 2006, 311:622-627.

20. Auffan M, Rose J, Bottero JY, Lowry GV, Jolivet JP, Wiesner MR: Towards a definition of inorganic nanoparticles from an environmental, health and safety perspective. *Nat Nanotech* 2009, 4:634-641.

21. Nel A, Madler T, Velegol D, Xia T, Hoek E, Somasundaran P, Klaessig F, Castranova V, Thompson M: Understanding biophysicochemical interactions at the nano-bio interface. *Nat Mater* 2009, 8:543-557.

22. Phenrat T, Kim HJ, Fagerlund F, Illangasekare T, Tilton RD, Lowry GV: Particle size distribution, concentration, and magnetic attraction affect transport of polymer-modified Fe^0 nanoparticles in sand columns. *Environ Sci Technol* 2009, 43:5079-5085.

23. Goon IY, Lai LMH, Lim M, Munroe P, Gooding JJ, Amal R: Fabrication and dispersion of gold-shell-protected magnetite nanoparticles: systematic control using polyethyleneimine. *Chem Mater* 2009, 21:673-681.

24. Takahashi K, Kato H, Saito T, Matsuyama S, Kinugasa S: Precise measurement of the size of nanoparticles by dynamic light scattering with uncertainty analysis. *Part Part Syst Charact* 2008, 25:31-38.

25. Goldburg WI: Dynamic light scattering. *Am J Phys* 1999, 67:1152-1160.

26. Chatterjee J, Haik Y, Chen CJ: Size dependent magnetic properties of iron oxide nanoparticles. *J Magn Magn Mater* 2003, 257:113-118.

27. DiPietro RS, Johnson HG, Bennett SP, Nummy TJ, Lewis LH: Determining magnetic nanoparticle size distributions from thermomagnetic measurements. *Appl Phys Lett* 2010, 96:222506.

28. Silva LP, Lacava ZGM, Buske N, Morais PC, Azevedo RB: Atomic force microscopy and transmission electron microscopy of biocompatible magnetic fluids: a comparative analysis. *J Nanopart Res* 2004, 6:209-213.

29. Dukhin AS, Goetz PJ: Acoustic and electroacoustic spectroscopy. *Langmuir* 1996, 12:4336-4344.

30. Chantrell RW, Wohlfarth EP: Rate dependent of the field-cooled magnetisation of a fine particle system. *Phys Status Solidi A* 1985, 91:619-626.

31. El-Hilo M, O'Grady K, Chantrell RW: Susceptibility phenomena in a fine particle system: I. Concentration dependence of peak. *J Magn Magn Mater* 1992, 114:295-306.

32. Jans H, Liu X, Austin L, Maes G, Huo Q: Dynamic light scattering as a powerful tool for gold nanoparticle bioconjugation and biomolecular binding studies. *Anal Chem* 2009, 81:9425-9432.

33. Ando K, Chiba A, Tanoue H: Uniaxial magnetic anisotropy of submicron MnAs ferromagnets in GaAs semiconductors. *Appl Phys Lett* 1998, 73:387.

34. Lacava LM, Lacava BM, Azevedo RB, Lacava ZGM, Buske N, Tronconi AL, Morais PC:Nanoparticles sizing: a comparative study using atomic force microscopy, transmission electron microscopy, and ferromagnetic resonance. *J Magn Magn Mater* 2001, 225:79-83.

35. Dukhin AS, Goetz PJ, Fang X, Somasundaran P: Monitoring nanoparticles in the presence of larger particles in liquids using acoustics and electron microscopy. *J Colloid Interface Sci* 2010, 342:18-25.

36. Van de Hulst HC: *Light Scattering by Small Particles*. New York: Dover Publications; 1981.

37. Hiemenz PC, Rajagopalan R: *Principles of Colloid and Surface Chemistry*. 3rd edition. New York: Marcel Dekker; 1997.

38. Berne BJ, Pecora R: *Dynamic Light Scattering: With Applications to Chemistry, Biology and Physics*. New York: Dover Publications; 2000.

39. He F, Zhao D: Manipulating the size and dispersibility of zerovalent iron nanoparticles by use of carboxymethyl cellulose stabilizers. *Environ Sci Technol* 2007, 41:6216-6221.

40. Tiraferri A, Chen KL, Sethi R, Elimelech M: Reduced aggregation and sedimentation of zero valent iron nanoparticles in the presence of guar gum. *J Colloid Interface Sci* 2008, 324:71-79.

41. Saleh N, Phenrat T, Sirk K, Dufour B, Ok J, Sarbu T, Matyjaszewski K, Tilton RD, Lowry GV:Adsorbed triblock copolymer deliver reactive iron nanoparticles to the oil/water interface. *Nano Lett* 2005, 5:2489-2494.

42. Vidal-Vidal J, Rivas J, López-Quintela MA: Synthesis of monodisperse maghemite nanoparticles by the microemulsion method. *Colloid Suface A: Physiochem Eng Aspects* 2006, 288:44-51.

43. Babič M, Horák D, Jendelová P, Glogarová K, Herynek V, Trchová M, Likavčannová K, Lesny P, Pollert E, Hájek M, Syková E: Poly(N, N-dimethylacrylamide)-coated maghemite nanoparticles for stem cell labelling. *Bioconjugate Chem* 2009, 20:283-294.

44. Kaufner L, Cartier R, Wüstneck R, Fichtner I, Pietschmann S, Bruhn H, Schütt D, Thünemann AF, Pison U: Poly(ethylene oxide)-block-poly(glutamic acid) coated maghemite nanoparticles: in vitro characterization and in vivo behavior. *Nanotechnology* 2007, 18:115710.

45. Thünemann AF, Schütt D, Kaufner L, Pison U, Möhwald H: Maghemite nanoparticles protectively coated with poly(ethyleneimine) and poly(ethylene oxide)-block-poly(glutamic acid). *Langmuir* 2006, 22:2351-2357.

46. Flesch C, Bourgeat-Lami E, Mornet S, Duguet E, Delaite C, Dumas P: Synthesis of colloidal superparamagnetic nanocomposites by grafting poly(ε-caprolactone) from the surface of organosilane-modified maghemite nanoparticles. *J Polym Sci A1* 2005, 43:3221-3231.

47. Nitin N, LaConte LEW, Zurkiya O, Hu X, Bao G: Functionalization and peptide-based delivery of magnetic nanoparticles as an intracellular MRI contrast agent. *J Biol Inorg Chem* 2004, 9:706-712.

48. Thompson Mefford O, Vadala ML, Goff JD, Carroll MRJ, Mejia-Ariza R, Caba BL, St Pierre TG, Woodward RC, Davis RM, Riffle JS: Stability of polydimethysiloxane-magnetite nanoparticle dispersions against flocculation: interparticle interactions of polydisperse materials. *Langmuir* 2008, 24:5060-5069.

49. Jain TK, Morales MA, Sahoo SK, Leslie-Pelecky DL, Labhasetwar V: Iron oxide nanoparticles for sustained delivery of anticancer agents. *Mol Pharmaceutics* 2005, 2:194-205.

50. Arsianti M, Lim M, Lou SN, Goon IY, Marquis CP, Amal R: Bi-functional gold-coated magnetite composites with improved biocompatibility. *J Colloid Interface Sci* 2011, 354:536-545.

51. Xie J, Xu C, Kohler N, Hou Y, Sun S: Controlled PEGylation of monodispersed Fe_3O_4 nanoparticles for reduced non-specific uptake by macrophage cells. *Adv Mater* 2007, 19:3163-3166.

52. Wan J, Cai W, Meng X, Liu E: Monodisperse water-soluble magnetite nanoparticles prepared by polyol process for high-performance magnetic resonance imaging. *Chem Commun* 2007, 5004-5006.

53. Narain R, Gonzales M, Hoffman AS, Stayton PS, Krishnan KM: Synthesis of monodisperse biotinylated p(NIPAAm)-coated iron oxide magnetic nanoparticles and their bioconjugation to streptavidin. *Langmuir* 2007, 23:6299-6304.

54. Gonzales M, Krishnan KM: Phase transfer of highly monodisperse iron oxide nanocrystals with Pluronic F127 for biomedical applications. *J Magn Magn Mater* 2007, 311:59-62.

55. Yeap SW, Ahmad AL, Ooi BS, Lim JK: Electrosteric stabilization and its role in cooperative magnetophoresis of colloidal magnetic nanoparticles. *Langmuir* 2012, 28:14878-14891.

56. Lim JK, Derek CJC, Jalak SA, Toh PY: Mat Yasin NH, Ng BW, Ahmad AL: rapid magnetophoretic separation of microalgae. *Small* 2012, 8:1683-1692.

57. Taylor RM, Huber DL, Monson TC, Ali AMS, Bisoffi M, Sillerud LO: Multifunctional iron platinum stealth immunomicelles: targeted detection of human prostate cancer cells using both fluorescence and magnetic resonance imaging. *J Nanopart Res* 2011, 13:4717-4729.

58. Ahmad T, Ramanujachary KV, Lofland SE, Ganguli AK: Magnetic and electrochemical properties of nickel oxide nanoparticles obtained by the reverse-micellar route. *Solid State Sci* 2006, 8:425-430.

59. Horie M, Fukui H, Nishio K, Endoh S, Kato H, Fujita K, Miyauchi A, Nakamura A, Shichiri M, Ishida N, Kinugasa S, Morimoto Y, Niki E, Yoshida Y, Iwahashi H: Evaluation of acute oxidative stress induced by nio nanoparticles in vivo and in vitro. *J Occup Health* 2011, 53:64-74.

60. Zhang Y, Chen Y, Westerhoff P, Hristovski K, Crittenden JC: Stability of commercial metal oxide nanoparticles in water. *Water Res* 2008, 42:2204-2212.

61. King S, Hyunh K, Tannenbaum R: Kinetics of nucleation, growth, and stabilization of cobalt oxide nanoclusters. *J Phys Chem B* 2003, 107:12097-12104.

62. Baldi G, Bonacchi D, Franchini MC, Gentili D, Lorenzi G, Ricci A, Ravagli C: Synthesis and coating of cobalt ferrite nanoparticles: a first step toward the obtainment of new magnetic nanocarriers. *Langmuir* 2007, 23:4026-4028.

63. Min GK, Bevan MA, Prieve DC, Patterson GD: Light scattering characterization of polystyrene latex with and without adsorbed polymer. *Colloids Surf A* 2002, 202:9-21.

64. Koppel DE: Analysis of macromolecular polydispersity in intensity correlation spectroscopy: the method of cumulants. *J Chem Phys* 1972, 57:4814-4820.

65. Lim JK, Majetich SA, Tilton RD: Stabilization of superparamagnetic iron oxide-gold shell nanoparticles in high ionic strength media. *Langmuir* 2009, 25:13384-13393.

66. Zhang L, He R, Gu HC: Oleic acid coating on the monodisperse magnetite nanoparticles. *Appl Surf Sci* 2006, 253:2611-2617.

67. Wang Z, Wen XD, Hoffmann R, Son JS, Li R, Fang CC, Smilgies DM, Hyeon TH:Reconstructing a solid-solid phase transformation pathway in CdSe nanosheets with associated soft ligands. *Proc Natl Acad Sci USA* 2010, 107:17119-17124.

68. Gittings MR, Saville DA: The determination of hydrodynamic size and zeta potential from electrophoretic mobility and light scattering measurements. *Colloid Suface A: Physiochem Eng Aspects* 1998, 141:111-117.

69. Elimelech M, Gregory J, Jia X, Williams RA: *Particle Deposition and Aggregation: Measurement, Modeling and Simulation.* Stoneham: Butterworth-Heinemann; 1998.

70. Wiogo HTR, Lim M, Bulmus V, Yun J, Amal R: Stabilization of magnetic iron oxide nanoparticles in biological media by fetal bovine serum (FBS). *Langmuir* 2011, 27:843-850.

71. Donselaar LN, Philipse AP: Interactions between silica colloids with magnetite cores: diffusion sedimentation and light scattering. *J Colloid Interface Sci* 1999, 212:14-23.

72. Golas PL, Lowry GV, Matyjaszewski K, Tilton RD: Comparative study of polymeric stabilizers for magnetite nanoparticles using ATRP. *Langmuir* 2010, 26:16890-16900.

73.   Phenrat T, Saleh N, Sirk K, Tilton RD, Lowry GV: Aggregation and sedimentation of aqueous nanoscale zerovalent iron dispersion. *Environ Sci Technol* 2007, 41:284-290.

74.   Cuevas GDL, Faraudo J, Camacho J: Low-gradient magnetophoresis through field-induced reversible aggregation. *J Phys Chem C* 2008, 112:945-950.

75.   Andreu JS, Camacho J, Faraudo J: Aggregation of superparamagnetic colloids in magnetic field: the quest for the equilibrium state. *Soft Matter* 2011, 7:2336-2339.

76.   Ditsch A, Lindenmann S, Laibinis PE, Wang DIC, Hatton TA: High-gradient magnetic separation of magnetic nanoclusters. *Ind Eng Chem Res* 2005, 44:6824-6836.

77.   Yeap SP, Toh PY, Ahmad AL, Low SC, Majetich SA, Lim JK: Colloidal stability and magnetophoresis of gold-coated iron oxide nanorods in biological media. *J Phys Chem C* 2012, 116:22561-22569.

78.   Shen L, Stachowiak A, Fateen SEK, Laibinis PE, Hatton TA: Structure of alkanoic acid stabilized magnetic fluids. A small-angle neutron and light scattering analysis. *Langmuir* 2001, 17:288-299.

79.   Lehner D, Lindner H, Glatter O: Determination of the translational and rotational diffusion coefficients of rodlike particles using depolarized dynamic light scattering. *Langmuir* 2000, 16:1689-1695.

80.   Nath S, Kaittanis C, Ramachandran V, Dalal NS, Perez JM: Synthesis, magnetic characterization, and sensing applications of novel dextran-coated iron oxide nanorods. *Chem Mater* 2009, 21:1761-1767.

81.   Lim JK, Tan DX, Lanni F, Tilton RD, Majetich SA: Optical imaging and magnetophoresis of nanorods. *J Magn Magn Mater* 2009, 321:1557-1562.

82.   Broersma S: Rotational diffusion constant of a cylindrical particle. *J Chem Phys* 1960, 32:1626.

83.   Broersma S: Viscous force and torque constants for a cylinder. *J Chem Phys* 1981, 74:6989.

84.   Vasanthi R, Bhattacharyya S, Bagchi B: Anisotropic diffusion of spheroids in liquids: slow orientational relaxation of the oblates. *J Chem Phys* 2002, 116:1092.

85. Phalakornkul JK, Gast AP, Pecora R: Rotational and translational dynamics of rodlike polymers: a combined transient electric birefringence and dynamic light scattering study. *Macromolecules* 1999, 32:3122-3135.

86. Farrell D, Dennis CL, Lim JK, Majetich SA: Optical and electron microscopy studies of Schiller layer formation and structure. *J Colloid Interface Sci* 2009, 331:394-400.

87. Fang XL, Li Y, Chen C, Kuang Q, Gao XZ, Xie ZX, Xie SY, Huang RB, Zheng LS: pH-induced simultaneous synthesis and self-assembly of 3D layered β-FeOOH nanorods. *Langmuir* 2010, 26:2745-2750.

Calcium Orthophosphate Coatings, Films and Layers

Sergey V Dorozhkin

Kudrinskaja sq. 1-155, Moscow, 123242, Russia

ABSTRACT

In surgical disciplines, where bones have to be repaired, augmented or improved, bone substitutes are essential. Therefore, an interest has dramatically increased in application of synthetic bone grafts. As various interactions among cells, surrounding tissues and implanted biomaterials always occur at the interfaces, the surface properties of the implants are of the paramount importance in determining both the biological response to implants and the material response to the physiological conditions. Hence, a surface engineering is aimed to modify both the biomaterials, themselves, and biological responses through introducing desirable changes to the surface properties of the implants but still maintaining their bulk mechanical properties. To fulfill these requirements, a special class of artificial bone grafts has been introduced in 1976. It is composed of various mechanically stable (therefore, suitable for load bearing applications) biomaterials and/or bio-devices with calcium orthophosphate coatings, films and layers on their surfaces to both improve interactions with the surrounding tissues and provide an adequate bonding to bones. Many production techniques of calcium orthophosphate coatings, films and layers have been already invented and new promising techniques are continuously investigated. These specialized coatings, films and layers used to improve the surface properties of various types of artificial implants are the topic of this review.

REVIEW

Introduction

All available materials have the specific characteristics of their own, namely, some of them are corrosive or biologically incompatible; some are sensitive to light or oxidation; some are hydrophilic or hydrophobic in nature, *etc.* Due to these reasons, various approaches have been already developed to modify the basic properties of diverse materials, and applying surface coatings, films or layers is a choice of option to solve some problems in a conventional form. For the particular case of artificial bone grafts, synthetic materials which are to be used in biological environments must display an adequacy of both their surface and bulk characteristics in order to fulfill the dual requirements of biocompatibility and suitable mechanical properties for the given application. Otherwise, due to a poor biocompatibility of improper compounds, fibrous tissues always encapsulate the implants made from such materials, which prolong the healing time. Considering that surface is always the first part of any insert that interacts with the host, various types of surface modifications have been developed to enhance biocompatibility and osteoconductivity of the implants (Ruckenstein and Gourisankar 1986).

On the other hand, it is well known that, due to the great chemical similarity to the inorganic part of bones and teeth of mammals, calcium orthophosphates (listed in Table 1) appear to be very friendly substances for the *in vivo* applications (Dorozhkin 2009, 2011; LeGeros 1991; Elliott 1994; Brown and Constantz 1994; Amjad 1997; Brès and Hardouin 1998; Chow and Eanes 2001; Hughes et al2002; Dorozhkin 2012). However, since calcium orthophosphate bulk materials have a ceramic nature, they are mechanically weak (brittle); therefore, they cannot be subjected to the physiological loads as encountered in human skeletons, other than compressive ones. Therefore, for many years, the clinical applications of calcium orthophosphates alone have been largely limited to non-load bearing parts of the skeleton due to their inferior mechanical properties. Attempting to combine the advantages of various materials, which is one of the major innovations over the last approximately 40 years, researchers started to deposit biocompatible calcium orthophosphates onto the surface of mechanically strong but bio-inert or bio-tolerant materials (Ong and Chan 1999; de Groot et al.1998; Campbell 2003; Kokubo 2008). For example, metallic implants are encountered in endoprosthesis (such as total hip joint replacements) and artificial teeth sockets because the requirements for a sufficient mechanical stability necessitate the use of a metallic body for such devices. As metals do not undergo bone bonding, *i.e.*, do not form mechanically stable links between the implant and bone tissues, they are coated by calcium orthophosphates exhibiting the bone-bonding ability between the metal and bone. After being implanted, calcium orthophosphate coatings, films and layers might be replaced by autologous bone because such coatings, films and layers participate in bone remodeling responses similar to natural bones (Ong and Chan 1999; Onoki and Hashida 2006; Kobayashi et al. 2007; Epinette and Geesink 1995; Willmann 1999; Schliephake et al.2006; Kokubo et al. 2003; Habibovic et al. 2005; Hahn et al. 2009). Minimal requirements for HA coatings, films or layers (Table 2) have first been described in 1992 in the Food and Drug Administration (FDA) guidelines (Callahan et al. 1994), as well as a little bit later in the ISO standards (1996). Afterwards, the FDA guidelines were updated in 1997 (U.S. FDA 1995), while the ISO standards were updated in (ISO 2000) and (ISO 2008).

Table 1: Existing calcium orthophosphates and their major properties (Doro-zhkin 2009, 2011)

Ca/P molar ratio	Compound	Formula	Solubility at 25°C -(-log(K_s))	Solubility at 25°C (g/L)	pH stability range in aqueous solutions at 25°C
0.5	Monocalcium phosphate monohydrate (MCPM)	$Ca(H_2PO_4)_2 \cdot H_2O$	1.14	approximately 18	0.0 to 2.0
0.5	Monocalcium phosphate anhydrous (MCPA or MCP)	$Ca(H_2PO_4)_2$	1.14	approximately 17	[a]
1.0	Dicalcium phosphate dihydrate (DCPD), mineral brushite	$CaHPO_4 \cdot 2H_2O$	6.59	approximately 0.088	2.0 to 6.0
1.0	Dicalcium phosphate anhydrous (DCPA or DCP), mineral monetite	$CaHPO_4$	6.90	approximately 0.048	[a]
1.33	Octacalcium phosphate (OCP)	$Ca_8(HPO_4)_2(PO_4)_4 \cdot 5H_2O$	96.6	approximately 0.0081	5.5 to 7.0
1.5	-Tricalcium phosphate (-TCP)	$-Ca_3(PO_4)_2$	25.5	approximately 0.0025	[b]
1.5	-Tricalcium phosphate (-TCP)	$-Ca_3(PO_4)_2$	28.9	approximately 0.0005	[b]
1.2 to 2.2	Amorphous calcium phosphates (ACP)	$Ca_xH_y(PO_4)z \cdot nH_2O$, $n = 3$ to 4.5%; 15 to 20% H_2O	[c]	[b]	approximately 5 to 12 [d]
1.5 to 1.67	Calcium-deficient hydroxyapatite (CDHA or Ca-def HA)[e]	$Ca_{10-x}(HPO_4)_x(PO_4)_{6-x}(OH)_{2-x}(0 < x < 1)$	approximately 85	approximately 0.0094	6.5 to 9.5
1.67	Hydroxyapatite (HA, HAp or OHAp)	$Ca_{10}(PO_4)_6(OH)_2$	116.8	approximately 0.0003	9.5 to 12
1.67	Fluorapatite (FA or FAp)	$Ca_{10}(PO_4)_6F_2$	120.0	approximately 0.0002	7 to 12
1.67	Oxyapatite (OA, OAp or OXA)[f]	$Ca_{10}(PO_4)_6O$	approximately 69	approximately 0.087	[b]

2.0	Tetracalcium phosphate (TTCP or TetCP), mineral hilgenstockite	$Ca_4(PO_4)_2O$	38 to 44	approximately 0.0007	b

[a] Stable at temperatures above 100°C.

[b] These compounds cannot be precipitated from aqueous solutions.

[c] Cannot be measured precisely. However, the following values were found: 25.7 ± 0.1 (pH = 7.40), 29.9 ± 0.1 (pH = 6.00), 32.7 ± 0.1 (pH = 5.28). The comparative extent of dissolution in acidic buffer is ACP >> α-TCP >> β-TCP > CDHA >> HA > FA.

[d] Always metastable.

[e] Occasionally, it is called "precipitated HA".

[f] Existence of OA remains questionable.

Dorozhkin

Dorozhkin *Progress in Biomaterials* 2012 1:1, doi: 10.1186/2194-0517-1-1

Table 2: FDA requirements for HA coatings (Callahan et al. 1994)

Properties	Specification
Thickness	Not specific
Crystallinity	62% minimum
Phase purity	95% minimum
Ca/P atomic ratio	1.67 to 1.76
Density	2.98 g/cm3
Heavy metals	< 50 ppm
Tensile strength	> 50.8 MPa
Shear strength	> 22 MPa
Abrasion	Not specific

Dorozhkin

Dorozhkin *Progress in Biomaterials* 2012 1:1, doi: 10.1186/2194-0517-1-1

General Knowledge on Coatings, Films and Layers

According to Wikipedia, the free encyclopedia, 'Coating is a covering that is applied to the surface of an object, usually referred to as the substrate. In many cases, coatings are applied to improve surface properties of the substrate, such as appearance, adhesion, wettability, corrosion resistance, wear resistance and scratch resistance. In other cases, in particular, in printing processes and semiconductor device fabrication (where the substrate is a wafer), the coating forms an essential part of the finished product.' (2012a). Obviously, all the aforementioned is also valid for films. A layer is another important definition. It is determined as a single thickness of some material covering a surface or forming an overlying part or segment.

Historically, involvement with coatings, films and layers dates to the metal ages of antiquity. Consider the ancient craft of gold beating and gilding, which has been practiced continuously for, at least, 4 millennia. The Egyptians appear to have been the earliest practitioners of this art. Many magnificent examples of statuary, royal crowns and coffin cases that have survived intact attest to the level of skills achieved. For example, leaf samples from Luxor dating to the Eighteenth Dynasty (1567 to 1320 BC) appear to be approximately 0.3-µm thick. Such leaves were carefully applied and bonded to smoothed wax or resin-coated wood surfaces in a mechanical (cold) gilding process to create the earliest coatings (Ohring 2002). Concerning the subject of this review, to the best of my findings, the first research paper on calcium orthophosphate coatings was published in 1976 (Sudo et al. 1976).

In spite of the fact that the technology of coatings, films and layers appears to be simultaneously one of the oldest arts and one of the newest sciences, the distinction among the coatings, films and layers is not well established yet; moreover, it may vary depending on the field of science and/or technology. For example, in food industry, the following statement has been published: 'An edible coating (EC) is a thin layer of edible material formed as a coating on a food product, while an edible film (EF) is a preformed, thin layer, made of edible material, which once formed can be placed on or between food components (McHugh 2000). The main difference between these food systems is that the ECs are applied in liquid form on the food, usually by immersing the product in a solution-generating substance formed by the structural matrix (carbohydrate, protein, lipid or multi-component mixture), and EFs are first molded as solid sheets, which are then applied as a wrapping on the food product'. (Falguera et al. 2011). To clarify this topic further, an extensive search in the scientific databases (Scopus, ISI Web of Knowledge) has been performed, and a great number of fixed collocations have been revealed. For example, according to Scopus (as of May 2012), a combination of words 'wear-protecting + coating' in the publication titles is used more frequently if compared with that of 'wear-protecting + film' (75 and 11 publications, respectively). On the contrary, a combination of words 'ferroelectric + film' in the publication titles is used much more frequently if compared with that of 'ferroelectric + coating' (5,861 and 28 publications, respectively). Concerning the subject of current review, a combination of words 'apatite + coating' is found in the titles of 2,635 publications, while those of 'apatite + film' and 'apatite + layer' are found in the titles of 427 and 370 publications, respectively. A similar correlation is valid for the combinations of words 'calcium + phosphate + coating', 'calcium + phosphate + film' and 'calcium + phosphate + layer' (they are found in the titles of 737, 138 and 149 publications, respectively). Therefore, both HA and all other calcium orthophosphates are most commonly associated with coatings. Perhaps, the aforementioned facts might be just a matter of terminology or even a habit for each particular sub-direction of science and technology.

Now it is necessary to classify various types of coatings, films and layers. In general, many possibilities are available. For example, they might be classified according to their structural material, such as metallic, polymeric, ceramic or composite coatings, films and layers. Furthermore, they might be classified according to their properties, such as biodegradability, edibility, transparency, reflectivity, conductivity, hardness, porosity, solubility, permeability, etc., as well as by the adhesion strength to various substrates. Besides, using a formation approach, all coatings, films and layers can be divided into two big categories: i) conversion ones, which are formed by reaction products of the base material (for example, formation of an oxide layer by surface oxidation) and ii) deposited ones. In turn, the deposited coatings, films and layers might be further classified according to the deposition techniques (Table 3). More to the point, since coatings and films may consist of either one or many individually deposited layers, all of them might be divided into monolayer coatings and films, and multilayer ones. While the former ones are produced by a single stage, the latter ones are produced by layer-by-layer deposition techniques. Furthermore, the individual layers of the multilayer coatings and films might be both indistinguishable from each other (in this case, the multilayer coatings and films behave as a thick monolayer) and distinguishable from each other. In the latter case, there might be an opportunity (sometimes, only hypothetical) to remove one or several individual layers from the surface, making coatings and films thinner. Finally yet importantly, all types of layers, coatings and films might be thin or thick. These terms appear to be relative, and the distinction between them is not well determined either; furthermore, it depends on the specific application. Nevertheless, in general, researchers consider a thin layer, film or coating as one ranging from fractions of a nanometer to several micrometers in thickness. Therefore, a thick layer, film or coating has thickness exceeding several micrometers. Interestingly that, according to the aforementioned scientific databases, all types of coatings, films and layers are much more often 'thin' than 'thick', namely, according to Scopus (as of May 2012), a combination of words 'thin + coating' in publication titles is used more frequently if compared with that of 'thick + coating' (2,608 and 468 publications, respectively). Similarly, a combination of words 'thin + film' in publication titles is used much more frequently if compared with that of 'thick + film' (144,106 and 7,077 publications, respectively), and a combination of words 'thin + layer' in publication titles is used much more frequently if compared with that of 'thick + layer' (26,603 and 1,443 publications, respectively). Concerning the physical state of the precursor materials, layers, coatings and films may be applied as liquids, gasses or solids, which might be used as still another classification type. The quality of coatings, films and layers is usually assessed by measuring their porosity, chemical composition, homogeneity, macro- and micro-hardness, bond strength and surface roughness.

Table 3: Various techniques to deposit bio-resorbable coatings, films and layers of calcium orthophosphates on metal implants (Sun et al. 2001; Yang et al. 2005; Narayanan et al. 2010)

Technique	Thickness	Advantages	Disadvantages
Thermal spraying	30 to 200 µm	High deposition rates; low cost	Line of sight technique; high temperatures induce decomposition; rapid cooling produces amorphous coatings; high temperatures prevent from simultaneous incorporation of biological agents
Plasma spraying	30 to 200 µm	High deposition rates; improved wear and corrosion resistance and biocompatibility	Line of sight technique; high temperatures induce decomposition; rapid cooling produces amorphous coatings; high temperatures prevent from simultaneous incorporation of biological agents
Magnetron sputtering	0.5 to 3 µm	Uniform coating thickness on flat substrates; high purity and high adhesion; dense pore-free coatings; excellent coverage of steps and small features; ability to coat heat-sensitive substrates	Line of sight technique; expensive; low deposition rates; produces amorphous coatings; high temperatures prevent from simultaneous incorporation of biological agents

Pulsed laser deposition (laser ablation)	0.05 to 5 μm	Coatings with crystalline and amorphous phases; dense and porous coatings; high adhesive strength	Line of sight technique; expensive; high temperatures prevent from simultaneous incorporation of biological agents
Ion beam deposition	0.05 to 1 μm	Uniform coating thickness; high adhesive strength	Line of sight technique; expensive; produces amorphous coatings
Dynamic mixing method	0.05 to 1.3 μm	High adhesive strength	Line of sight technique; expensive; produces amorphous coatings
Dip and spin coating	2 μm to 0.5 mm	Inexpensive; coatings applied quickly; can coat complex substrates	Requires high sintering temperatures; thermal expansion mismatch
Sol–gel technique	<1 μm	Can coat complex shapes; low processing temperatures; thin coatings; inexpensive process; can incorporate biological molecules	Some processes require controlled atmosphere processing; expensive raw materials
Electrophoretic deposition	0.1 to 2.0 mm	Uniform coating thickness; rapid deposition rates; can coat complex substrates; can incorporate biological molecules	Difficult to produce crack-free coatings; requires high sintering temperatures

Electrochemical (cathodic) deposition	0.05 to 0.5 mm	Good shape conformity; room temperature process; uniform coating thickness; short processing times; can incorporate biological molecules	Sometimes stressed coatings are produced, leading to their poor adhesion with substrate; requires good control of electrolyte parameters
Biomimetic process	< 30 μm	Low processing temperatures; can form bonelike apatite; can coat complex shapes; can incorporate biological molecules	Time consuming; requires replenishment and a pH constancy of the simulating solutions (HBSS, SBF, etc.)
Hot isostatic pressing	0.2 to 2.0 μm	Produces dense coatings	Cannot coat complex substrates; high temperature required; thermal expansion mismatch; elastic property differences; expensive; removal/ interaction of encapsulation material; high temperatures prevent from simultaneous incorporation of biological agents
Micro-arc oxidation	3 to 20 μm	Simple, economical and environmentally friendly coating technique, suitable for coating of complex geometries	Except of calcium orthophosphates, coatings always contain admixture phases

HBSS, Hank's balanced salt solution; SBF, simulated body fluid.
Dorozhkin

Dorozhkin *Progress in Biomaterials* 2012 1:1, doi: 10.1186/2194-0517-1-1

Further, one should mention on the reasons why people apply layers, coatings and/or films to the surface of various materials. They are various, for example:

- The core contains a material, which is toxic, provokes adverse responses, allergic reactions, etc., or has a bitter taste, an unpleasant odor, etc.;

- Layers, coatings and/or films protect the core material from the surroundings to improve its stability and shelf life;

- Layers, coatings and/or films develop the mechanical integrity, which means that coated products are more resistant to mishandling (abrasion, attrition, etc.);

- To modify surface properties of the core, such as biocompatibility, light reflection, electrical conductivity, color, etc.;

- Decoration (in the cases, when the core alone is inelegant);

- The core contains a material, which migrates easily to stain hands, clothes and other objects;

- To modify the release profile of active components, e.g., drugs, from the core.

Reason numbers 1, 2, 3, 4 and 7 appear to be applicable to the biomedical field in general, while reason numbers. 1, 2 and 4 are relevant to the subject of this review.

To conclude this section, one should note that, in a certain sense, all types of coated materials resemble the functionally graded ones but with an extremely high gradient in both the composition and properties at the core/coating interface.

Preparation

Brief Knowledge on the Important Pre- and Post-Deposition Procedures

Due to the unfavorable mechanical properties of bulk calcium orthophosphate bioceramics, an extensive research has been focused on the development of calcium orthophosphate coatings, films and layers on the surfaces of various materials. Various deposition techniques have been already proposed, which are discussed below. The major advantages and disadvantages of the available deposition techniques have been summarized in Table 3 (Sun et al. 2001; Yang et al. 2005; Narayanan et al. 2010). However, in the vast majority of the cases, prior to be coated, an object (a substrate) needs to be prepared for coating. This normally consists of some kind of cleaning (e.g., ultrasonically in an acetone or ethanol bath to remove dirt, oil and other contaminants adhering to the surface) and may include etching, tarnishing, grounding and/or application of a conversion coating (Chou and Chang 2001). Besides, various types of physical modifications of the surface, such as sand- (Cao et al. 2010a) or grit-blasting and polishing, as well as wetting or drying procedures might be applied as well. Furthermore, after calcium orthophosphate coatings, films and/or layers have been formed, various types of post-deposition treatments might be also necessary to improve their properties. For example, post-deposition heat treatment (annealing) of calcium orthophosphates leads to conversion of the deposited amorphous and non-apatite phases into HA with simultaneous increasing of coating crystallinity, enhancing corrosion resistance, as well as reducing the residual stress (Ji and Marquis 1993; Ong and Lucas 1994; Yoshinari et al. 1997; Erkmen 1999; Burgess et al.1999; Sridhar et al. 2003; Yang et al. 2003a; Lee et al. 2005a; Johnson et al. 2006; Yang et al.2009a). The annealing can be done by various ways, including laser treatment (Cannillo et al. 2009) or electric polarization in alkaline solutions (Huang et al. 2009). Furthermore, the presence of water during the post-deposition heat treatment also plays an important role in this conversion (Yang et al.2009a; Cao et al. 1996; Yang et al. 2003b), namely, in comparison to heat treatments at 450°C in dry conditions, the presence of water vapor resulted in a significant increase in coating crystallinity (Yang et al. 2003b). Similar positive effect of hot water was obtained in other studies (Yang et al.2009a; Saju et al. 2009; Ozeki et al. 2010) in which post-deposition hydrothermal treatment at 100°C to 170°C was used (Figure 1).

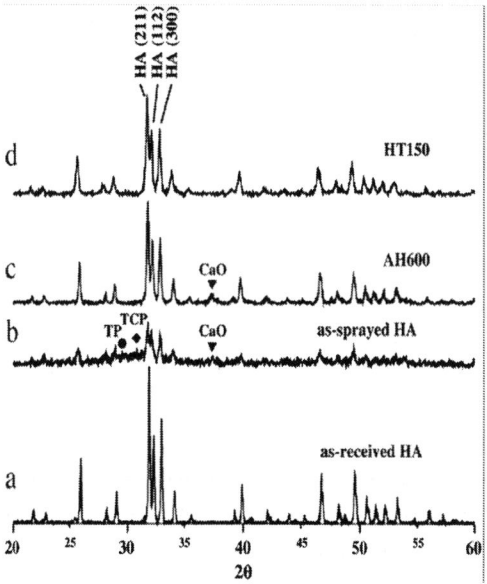

Figure 1: The XRD patterns. (a) An initial HA powder, (b) as sprayed HA coatings, (c) air heat-treated at 600°C HA coatings (AH600) and (d) hydrothermally treated at 150°C HA coatings (HT150). It is easily seen that both the crystallinity and phase purity of poorly crystalline HA coatings increased after both heat and hydrothermal treatments. Reprinted from Yang et al. (2009a) with permission.

Thermal Spraying Techniques

Thermal spraying is the process in which melted, softened or heated materials are sprayed onto a surface to be deposited on it. A feedstock with a coating material or a coating precursor might be heated by various ways, such as a high temperature flame or a plasma jet, by means of which thermal spraying is classified into flame spraying and plasma spraying. The principal difference between these two techniques is the maximum temperature achievable. In general, thermal spraying provides thick (from approximately 20 μm to several millimeters, depending on the process and feedstock) coatings, films and layers over a large area at high deposition rate, as compared to other coating processes such as electroplating, physical and chemical vapor deposition (Table 3). Coating materials are fed in powder or wire form, heated to a molten or semi-molten state and accelerated towards substrates in the form of micrometer-size particles. Resulting coatings or films are formed by a continuous buildup of successive layers of liquid droplets, softened material domains and hard particles. For all types of thermal spraying techniques, the quality of coatings, layers and films is generally increased with increasing particle velocities (Fauchais et al. 2001).

Since thermal spraying occurs at very high temperatures, the substrates are heated up as well. In some cases, this might result in phase transformation and recrystallization of the near surface zones. For example, a martensitic transformation and recrystallization was found to occur in near surface of a low-modulus Ti-24Nb-4Zr-7.9Sn alloy substrate after application of a plasma-sprayed HA coating. Both phenomena were attributed to the combination of temperature with cooling process (Zhao et al.2011). Certainly, such phenomena introduce additional ambiguities to the mechanical and adhesive properties of the deposited coatings, films and layers.

Plasma Spraying

Plasma is often referred to as the fourth state of matter, as it differs from solid, liquid and gaseous states, and does not obey the classical physical and thermodynamic laws. Plasmas are used in many different processing techniques, for example, for modification and activation of various surfaces. Much research is currently being done to understand and control them (Freidberg 2007).

According to de Groot et al. (1998), a plasma spraying technique was discovered accidentally in 1970 by a student, who used the equipment to study melted and rapidly solidified aluminum oxide coatings on a metal substrate (Herman 1988). In plasma spraying, a material to be deposited (feedstock) - typically as a powder, sometimes as a liquid, suspension or wire - is introduced into a plasma jet, emanating from a plasma torch (other names: plasma arc or plasma gun) because a stream of gasses (usually, argon; however, helium, hydrogen or nitrogen might be used as well) passes through this torch. The torch turns these gasses into ionized plasma of a very high temperature (up to approximately 20,000 K) and with a high speed of up to 400 m/s. In the jet, the material is either melted or heat softened, and these molten or softened droplets flatten and propel towards a substrate to be deposited on it. Appropriate cooling techniques keep the temperature of the substrate below 100°C to 150°C. Since the temperature of the plasma rapidly decreases as a function of distance, the droplets rapidly solidify and form deposits. Commonly, the deposits consist of a multitude of pancake-like lamellae called 'splats', formed by flattening of the liquid or softened droplets. They remain adherent to the substrate as coatings, films and layers. As the feedstock typically consists of powders with sizes from several micrometers to approximately 1 mm, the lamellae have thickness in the micrometer range and lateral dimension from several to hundreds of micrometers. That is why thick coatings, films or layers might be produced only. One pass of the plasma gun can produce a layer of about 5 to 15-lamellae thick. Once a layer has been applied to the whole substrate, the gun returns to the initial position and another layer is applied (Fauchais 2004). Typical current values that are used for spraying HA coatings range from 350 A (Cao et al. 1996) to 1,000 A (Quek et al. 1999). Good schematic setups of the plasma spraying process are available in literature (Narayanan et al. 2010; Fauchais 2004; Paital and Dahotre 2009a; Layrolle 2011; Surmenev 2012). A typical image of a plasma-sprayed HA coating is shown in Figure 2 (Layrolle 2011).

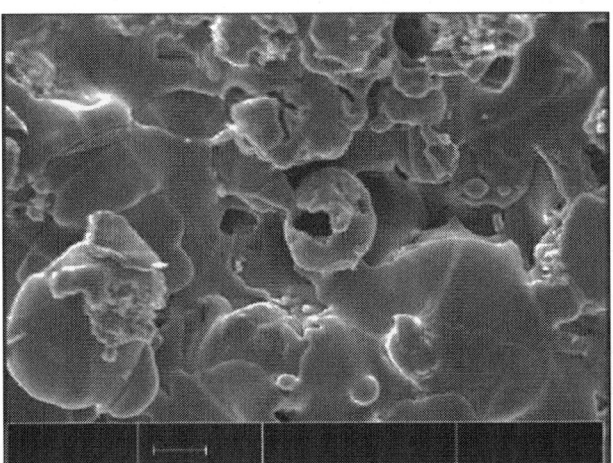

Figure 2: A scanning electron microscopy of a typical plasma sprayed HA coating on titanium implants. Bar is 10 μm. Reprinted from Layrolle (2011) with permission.

Depending on the experimental conditions, various sub-modifications of plasma spraying technique have been outlined, namely atmospheric plasma spraying (Heimann 2006), vacuum (or low-pressure) plasma spraying (Gledhill et al. 1999, 2001), powder plasma spraying, suspension plasma spraying (Jaworski et al. 2009; Gross and Saber-Samandari 2009; Podlesak et al. 2010), liquid (or solution) plasma spraying (Huang et al. 2010) and, gas plasma spraying (Morks and Kobayashi 2007; Wu et al.2009) techniques, and all of them are used to fabricate bioactive calcium orthophosphate-based coatings, films and layers. Such modifications have some specific advantages, e.g., they allow obtaining thinner coatings of 5 to 50 μm, which are a few times thinner than those obtained by dry powder processing (Table 3) (Surmenev 2012). In addition, there is a microplasma spraying technique (Dey and Mukhopadhyay 2010; Dey et al. 2011; Dey and Mukhopadhyay 2011), which is characterized by small dimensions, a low level (25 to 50 dB) of noise and hardly any dust, as well as a low power consumption. All of these make it possible to operate under normal workroom conditions. The process provides deposition of coatings, films and layers on small-sized parts and components, including those with fine sections, this being unachievable with any other methods. Due to a low heat input of the microplasma jet, overheating of the powder particles as well as excessive local overheating of the substrate is reduced. The mechanical properties of the coatings, films and layers are generally good (Dey and Mukhopadhyay 2010; Dey et al. 2011; Dey and Mukhopadhyay 2011). This makes it possible to widen the application scales of plasma spraying and produce different functional coatings.

It is important to stress that, among the deposited lamellae, there are small voids, such as pores, cracks and regions of incomplete bonding. Due to such inhomogeneous structure, the deposits can have properties significantly different from the initial bulk materials (Fauchais 2004). In addition, due to both very high processing temperatures (leading to dehydroxylation of HA and partial decomposition of any other material) followed by rapid solidification, various admixtures and metastable phases are usually present in the deposits. For example, in the case of plasma spraying of calcium orthophosphates, complicated mixtures of various phases (high temperature ACP, -TCP, -TCP, HA, OA, TTCP) with other compounds, such as calcium pyrophosphates, calcium metaphosphates, CaO, etc. are obtained (Cao et al. 2010a;

Zyman et al. 1993; Weng et al. 1993; Zyman et al. 1994; Tong et al. 1995; McPherson et al. 1995; Yang et al. 1995; Tong et al. 1998; Park et al. 1998; Gross et al. 1998a; Gross et al. 1998b; Heimann and Wirth 2006; Roy et al. 2011). Thus, the chemical and phase compositions of the final coatings, films and layers are dependent on the thermal history of the powder particles. This leads to variable solubility of the deposited coatings, films and layers, dictated by the amounts of more soluble phases, such as ACP. Furthermore, the distribution of by-product and metastable phases in the coatings, films and layers appears to be inhomogeneous. For example, the coating crystallinity was reported to be lower at the interface with the Ti substrate than at the surface of the coating. This happened because metals had a higher rate of thermal diffusivity than calcium orthophosphates and, thus, the cooling rate of the first layer was faster (Gross et al. 1998b). Besides, residual stresses in the plasma sprayed coatings, films and layers were found and measured (Valter et al. 1997; Tsui et al. 1998; Yang et al. 2000; 2003c; Yang and Chang 2005; Carradó 2010; Yang 2011).

A diagram for the formation of various phases during plasma spraying of HA coatings is presented in Figure 3 (Khor and Cheang 1993). According to the authors, if the outside skin of an HA particle is molten and the core remains un-molten, insufficient heat is transferred to melt the particle completely. This model is modified to a totally molten hydroxyl-rich core (with the stoichiometry of HA) with further changes depending on the heat transfer to the particle. The first condition depicts a molten droplet with a hydroxyl-depleted skin. The center containing the hydroxyl-rich molten material will crystallize upon deposition to form HA. The dehydroxylated region, which is exposed to the substrate upon droplet spreading, will form an ACP phase, but the area distant from the substrate will crystallize to form OA. OA requires smaller atomic rearrangements to occur for crystallization from a viscous melt, and, therefore, crystallizes in preference to a mixture of TCP and TTCP. Growth of HA will begin in the hydroxyl-rich core and will finally change to OA in response to the depleted hydroxyl concentration at the top of the lamellae (Figure 3, case (i)). If the molten particle flattens to an extent where the cooling rate is increased, then the entire particle becomes amorphous. Both TCP and TTCP are observed in greater quantities when a higher heat transfer to the particle prevails. If the heat dissipation is slow through the already-solidified amorphous and crystalline layers of the coating, TCP and TTCP can be nucleated at the top surface of the lamellae (Figure 3, case (ii)). The growth of TCP and TTCP may delay the growth of OA with the latent heat of fusion. With a high level of dehydroxylation in the molten particle, lesser amounts of HA or OA will form, and so the large volume of dehydroxylated material will then mostly contain TCP and TTCP. The growth mechanism may begin within the droplet, since a more fluid droplet facilitates faster diffusion. Calcium oxide is observed when even higher heating conditions are employed. In addition to being hydroxyl deficient, the outer shell of the molten particle also becomes phosphate deficient (Figure 3, case (iii)) (Narayanan et al.2010). In addition, a numerical simulation model of HA powder behavior in plasma jet was suggested (Dyshlovenko et al. 2004). Within this model, the authors created temperature fields inside an HA particle before impact and their transformation into crystal phases after rapid solidification and cooling (Figure 4).

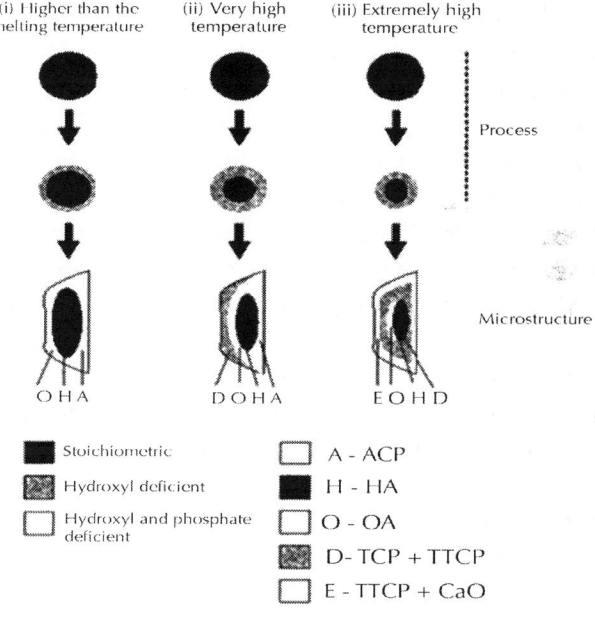

Figure 3: A proposed model for phase formation in the plasma sprayed HA coatings. The process stage depicts the various melt chemistries as a function of particle temperature. The microstructure depicts the different phases that can be formed in a lamella. Reprinted from Khor and Cheang (1993) with permission.

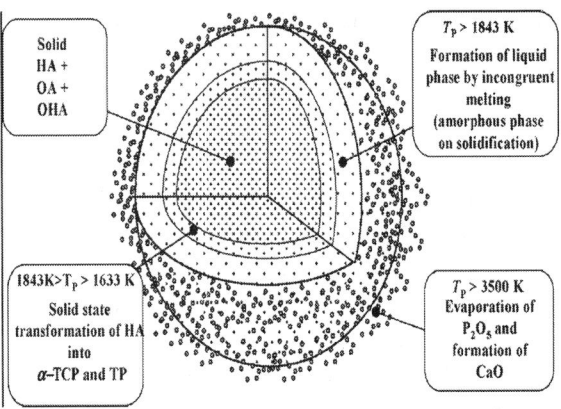

Figure 4: Temperature fields of HA powder particle at impact and assumed phase transformations. Reprinted from Dyshlovenko et al. (2004) with permission.

There are a large number of technological parameters that influence the interaction of the particles with the plasma jet and the substrate and, therefore, the properties of final coatings, films and layers. These parameters include feedstock type, plasma gas composition, flow rate, energy input, torch offset distance, substrate cooling, etc. (Cizek et al. 2007). Furthermore, due to the very high temperatures of plasmas, the aforementioned thermodynamic instability of calcium orthophosphates at such temperatures plays an important role in the final properties of the deposited coatings, films and layers. Ideally, only a thin outer layer of each powder particle should be heated to the molten plastic state, which unavoidably undergoes both chemical transformations and phase transitions. This plastic state is necessary to ensure dense and adhesive coatings but it should comprise just a negligible volume fraction of the particles. By choosing optimum relations among particle size, type of gas, speed of the plasma and cooling process of the coated surface, one obtains calcium orthophosphate coatings films and layers with the desired thickness and crystallinity (de Groot et al.1987; Cook et al. 1988; Stevenson et al. 1989; Wolke et al. 1992; Sun et al. 2003; Prymak et al.2004).

The dimensions of calcium orthophosphate particles were found to affect their melting characteristics within the plasma flame, namely, large particles undergo a lesser degree of melting in the plasma flame than small particles (Cheang and Khor 1995; Kweh et al. 2000). For example, during spraying of HA particles with dimensions exceeding approximately 55 µm they were found to remain crystalline and showed little or no melting during plasma spraying. HA particles with dimensions within 30 to 55 µm were partially melted and consisted of mixtures of crystalline and amorphous phases, while HA particles less than approximately 30 µm were fully melted and contained large amounts of ACP and traces of CaO (Cheang and Khor 1995). In another study, plasma sprayed HA particles of 20 to 45 µm in size were found to produce denser lamellar coatings than the coatings obtained by plasma spraying of 45 to 75 and 75 to 125-µm HA particles. Coatings formed from 20 to 45-µm sized HA particles did not show the presence of cavities but contained a flatter smoother surface profile as a result of neatly stacked disk-like splats, while coatings formed from 45 to 75 and 75 to 125-µm sized HA particles contained numerous un melted particles, cavities and macropores (Kweh et al. 2000). Interestingly, the coating roughness might be used as a measure of the melting degree of particles within the plasma flame, namely, when the particles reach a more fluid state within the plasma flame, they become less viscous and can be spread out to a greater degree on impact with the substrate. A smoother coating will result in this case. Partially melted particles will not be able to flatten on the coating surface. This situation will lead to large undulations and a rough coating (Gross and Babovic2002).

Further details on the plasma spraying technique are available in excellent reviews (Surmenev 2012).

High Velocity Oxy-Fuel Spraying

In 1990s, a new class of thermal spray processes called high velocity oxy-fuel (HVOF) spraying was developed (Oguchi et al. 1992; Sobolev and Guilemany 1996). A mixture of gaseous (hydrogen, methane, propane, propylene, acetylene, natural gas, etc.) or liquid (kerosene, etc.) fuel and oxygen is fed into a combustion chamber, where they are ignited and combusted continuously. The resultant hot gas at a pressure approximately 1 MPa emanates through a converging-diverging nozzle and travels through a straight section. The jet velocity (> 1,000 m/s) at the exit of the barrel exceeds the speed of sound. A powder feed stock is injected into the gas stream, which accelerates the powder up to 800 m/s. The stream of hot gas and powder is directed towards the surface to be coated. The powder partially melts in the stream and deposits upon the substrate. The resulting calcium orthophosphate coatings, layers and films have a low porosity and a high adhesion strength (Oguchi et al. 1992; Sobolev and Guilemany 1996; Haman et al. 1999; Li et al. 2000, 2002a; Khor et al.2003a, 2003b, 2004; Hasan and Stokes 2011).

Similar to the aforementioned results on plasma spraying, in the case of HVOF spraying, larger particles of calcium orthophosphates were also found to undergo a lesser degree of melting than smaller particles (Khor et al. 2003b), namely, cross-sectional SEM investigations of the sprayed HA particles of $50 \pm 10\,\mu m$ in sizes revealed that they were melt only partially from the surface, while those for HA particles of $30 \pm 10\,\mu m$ in sizes revealed that they were melt almost completely. The coating morphology shown in Figure 5 further reveals the influence of the melt state on grain size of the coatings. It clearly demonstrates the interface zone between the melted and un melted parts within a HA splat. It is noted that the HA grains located in un melted part are of far larger size than those in melted part, which states the influence of rapid cooling on grain growth during coating formation (Khor et al. 2003b). Furthermore, Raman spectroscopy qualitative inspection on the sprayed HA particles (partially melted) revealed that a thermal decomposition of HA occurred within the melted part rather than the unmelted zone (Khor et al. 2004). Therefore, to both achieve high crystallinity of the coatings and reduce the amount of admixture phases, the appropriate powder size together with the apt HVOF spray parameters must be carefully selected.

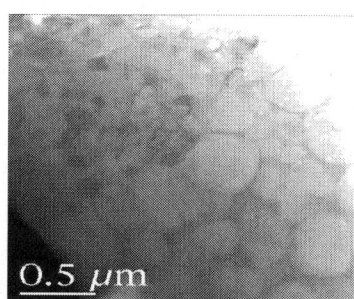

Figure 5: TEM image of as-sprayed HA coating. This show the interface between unmelted and un-melted parts within a HA splat and different grain size. Reprinted from Khor et al. (2003b) with permission.

Wet Techniques

As follows from the definition, all types of wet deposition techniques occur from either solutions or suspensions both aqueous and non-aqueous. Furthermore, all of them occur at moderate temperatures (Nijhuis et al. 2010). Depending on the solution pH, various calcium orthophosphates might be precipitated (Table 1) and, therefore, be deposited as coatings, films and layers. In general, the deposition kinetics depends on the solution supersaturation, concentration of the reagents, temperature, presence or absence of admixtures, nucleators, inhibitors, etc. As to the precipitation mechanism of calcium orthophosphates from aqueous solutions, this process appears to be rather complicated; for the biologically relevant calcium orthophosphates (OCP, CDHA and HA), the crystallization process occurs via formation of one or several intermediate and/or precursor phases, such as ACP, DCPD and/or OCP. The detailed description of the precipitation mechanisms of various calcium orthophosphates is beyond the scope of current review; the interested readers are referred to the special literature on the topic (Wang and Nancollas 2008; Wang et al. 2011).

For some types of the wet techniques, specific surface preparation techniques appear to be necessary. For example, if calcium orthophosphates need to be deposited on titanium or its alloys, a surface layer of hydrated titanium hydroxides should be created prior the deposition (Wang et al.2008a). This can be done by various oxidation techniques, such as alkali treatment (de Andrade et al.2002; Liang et al. 2003; Wang et al. 2000), oxidation in H_2O_2 (Wang et al. 2000), micro-arc oxidation (Song et al. 2004), pre-calcification in boiling $Ca(OH)_2$ solution (Wen et al. 1997; Chen et al. 2009a) or using water vapor treatment (Feng et al. 2002). Positive effects of the presence of hydrated silica (Li et al. 1994) and sodium (Pham et al. 2002) on the surface are known as well. Since the detailed description of the surface preparation of metals is beyond the scope of this review, the interested readers are referred to the special literature on the subject (Narayanan et al. 2010; Nanci et al. 1998; Liu et al. 2004; Rautray et al. 2010; Variola et al. 2011).

Electrophoretic Deposition

According to Wikipedia, the free encyclopedia. 'Electrophoretic deposition is a term for a broad range of industrial processes, which includes electrocoating, e-coating, cathodic electrodeposition and electrophoretic coating or electrophoretic painting.' (2012b). A characteristic feature of this process is that charged colloidal particles suspended in a liquid medium migrate under the influence of a direct current electric field (electrophoresis) and are deposited onto a conductive substrate of the opposite charge (Besra and Liu 2007).

Since electrophoretic deposition is designed to apply materials to any electrically conductive surface, it is used to achieve calcium orthophosphate coatings, layers or films on various metallic substrates only (Ducheyne et al. 1986, 1990; Zhitomirsky and Gal-Or 1997; Han et al. 1999a; Zhitomirsky 2000; Wei et al. 2001; Stoch et al. 2001; Wang et al. 2002; de Sena et al. 2002; Ma et al. 2003a; Mondragón-Cortez and Vargas-Gutiérrez 2004; Meng et al. 2006, 2008). This approach is especially useful for porous metallic structures. To create coatings, layers or films, calcium orthophosphate powders are suspended in water or other suitable liquids to produce a coating bath, followed by deposition onto a metallic surface. The proper dimensions of the particles to be deposited are very important because the particles must be fine enough to remain in suspension during the coating process. Electrophoretic deposition normally involves submerging a metallic substrate into a container or vessel, which holds the coating bath, and applying direct current electricity using electrodes, where the substrate is one of the electrodes (anode or cathode). An applied electric field is the driving force of the deposition (Besra and Liu 2007). Depending on the mode and sequence of voltage applied, electrophoretic deposition of calcium orthophosphates can be carried out at either constant (Meng et al. 2006) or dynamic (Meng et al. 2008) voltage.

After deposition, an object is normally rinsed off to remove the undeposited bath, followed by sintering in a high vacuum (10^{-6} to 10^{-7} Torr) at 850°C to 950°C (Besra and Liu 2007). The resulting coatings, layers or films consist of a number of calcium orthophosphate phases plus various random admixtures. For example, in the case of electrophoretically deposited CDHA coatings, the sintering results in their transformation to biphasic (HA + β-TCP) coatings (Han et al.

1999a). Their thicknesses can be varied by changing the electrical field strength and the deposition time. Further, at the coating/substrate interface various metal-phosphorus compounds might be formed due to mutual inter-diffusion of calcium orthophosphates and atoms of the metallic substrate. Unfortunately, due to densification during sintering, shrinkage and cracking of the coatings, layers or films can occur. In addition, thermal stresses induced by the differences in thermal expansion coefficients between the core and the coating during sintering and cooling can lead to cracking (de Groot et al. 1998).

The surface morphology of the electrophoretically deposited calcium orthophosphate coatings was found to depend on applied voltage (Mondragón-Cortez and Vargas-Gutiérrez 2004), deposition time (Mondragón-Cortez and Vargas-Gutiérrez 2004) and powder concentration (Meng et al. 2006), namely, at 200 V, the deposited particles had dimensions within 0.20 to 0.35 μm; at 400 V, the particle size range increased up to 0.35 to 0.80 μm and at 800 V, the particle size range increased up to 0.80 to 1.20 μm. Furthermore, increasing voltages resulted in increasing of the amount of deposited calcium orthophosphates. Besides, porous and roughened coatings were obtained at a higher electric field, while dense coatings of finer particle size were obtained at a lower electric field (Mondragón-Cortez and Vargas-Gutiérrez 2004). Similar effect was noticed for the deposition time: the shorter the time, the smaller particles were deposited (Mondragón-Cortez and Vargas-Gutiérrez2004). Concerning the powder concentration in suspensions, for a low HA concentration, the coatings were very rough, and a great level of agglomeration was noticed. At higher HA concentrations, the coatings became uniform and crack free, and there was less agglomeration. At very high concentrations of HA, many cracks were found (Meng et al. 2006). These results indicate that powder concentration, deposition time and applied potential have a significant effect on the coating morphology.

Interestingly, some specific types of calcium orthophosphate bioceramics might be prepared by electrophoretic deposition (Zhitomirsky 2000; Wang et al. 2002; Ma et al. 2003b). For example, hollow HA fibers of various diameters were fabricated (Zhitomirsky 2000). In the first step, submicron HA powders were electrophoretically deposited on individual carbon fibers, carbon fibers bundles and felts. Then, they were burned out and sintered to remove the carbon substrate and leave behind the corresponding ceramic replicas (Zhitomirsky

2000). Similarly, uniform HA tubes were prepared by electrophoretic deposition of HA powders on carbon rods by repeated depositions at room temperature (Wang et al. 2002). The repeated deposition process was necessary to produce thicker multilayered coatings with no surface cracks. The green bodies were then sintered under a range of temperatures varying from 1,150°C to 1,300°C to burn out carbon and obtain HA tubes (Wang et al.2002). Furthermore, porous calcium orthophosphate scaffolds were fabricated by electrophoretic deposition (Ma et al. 2003b).

To conclude, electrophoretically deposited calcium orthophosphate coatings on implants are commercially available. The examples include BIONIT® (DOT GmbH, Rostock, Germany) and BoneMaster® (BIOMET Corp., Warsaw, IN, USA) (Layrolle 2011). In addition, various modifications and hybrid technologies, such as plasma-assisted electrophoretic deposition (Nie et al. 2001) and a combination of micro-arc oxidation with electrophoresis (Nie et al. 2000) have been developed as well.

Electrochemical (Cathodic) Deposition

In electrochemical deposition of calcium orthophosphates, a supersaturated or a metastable aqueous electrolyte containing calcium and orthophosphate ions is used. Various electrochemical reactions occurring in the electrolyte near electrodes induce local pH increase, and thus, calcium orthophosphate crystals are nucleated and grow on the electrodes (Manso et al. 2000; Duan et al.2003; Lu et al. 2005). Obviously, only conductive materials might be coated by this technique. A typical setup includes a platinum electrode (anode) and a metallic implant (cathode) connected to a current generator. Since electrochemical deposition usually occurs on the negatively charged electrodes, in literature it is sometimes referred to as cathodic deposition (Zhao et al. 2003; Blackwood and Seah 2009; Roguska et al. 2011). The electrochemical reactions occurring with the ions during the deposition of calcium orthophosphates might be found in literature (Kuo and Yen2002; Yen and Lin 2002).

Since electrochemical deposition of calcium orthophosphate coatings, films and layers occurs from aqueous solutions, it is commonly performed at ambient conditions (Rossler et al. 2003; Lin et al.2003).

However, electrochemical deposition performed in an autoclave at 80°C to 200°C is also known (Ban and Maruno 1998). The process might be performed in various electrolytes, including SBF (Wang et al. 2004; Lopez-Heredia et al. 2007). A typical example of the deposited coating is shown in Figure 6 (Layrolle 2011). A coating thickness of less than 1 µm can be achieved. Reduction of the thickness leads to an increased resistance to delamination, which is observed frequently for thicker coatings (Peng et al. 2006). Electrochemical deposition of nano-sized crystals is also possible (Shirkhanzadeh 1998; Yousefpour et al. 2006; Narayanan et al. 2007, 2008a, 2008b, 2008c. Natural materials, such as shells, have been tested as the source of calcium to produce coatings by electrochemical deposition (Narayanan et al. 2006). Unfortunately, deposition of calcium orthophosphates requires a sufficient volume of electrolyte to surface ratio. Besides, approximately 20 ml of electrolyte is needed to coat 1 cm^2 of implant. Additionally, hydrogen gas production hampers the deposition due to formation of bubbles on the titanium surface, which results in non-uniform coatings (Layrolle 2011). In order to overcome the latter problem, a modulated electrochemical deposition technique has been proposed (Lin et al. 2003).

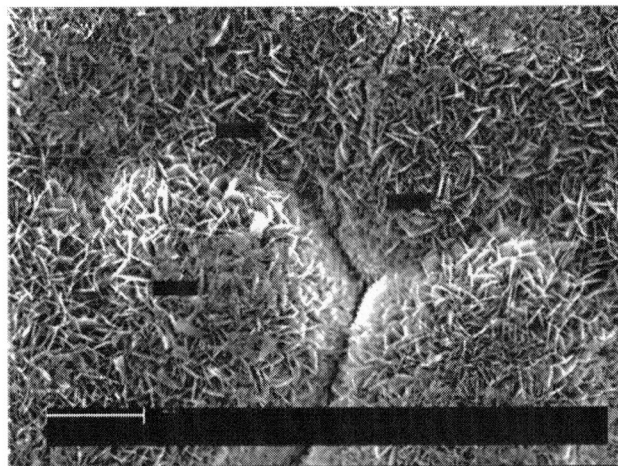

Figure 6: A scanning electron microscopy of a typical electrochemically deposited coating on titanium. Bar is 20 µm. Reprinted from Layrolle (2011) with permission.

According to the literature, nucleation of calcium orthophosphate crystals during the electrochemical deposition can occur either as instantaneous nucleation or as progressive nucleation (Eliaz and Eliyahu 2007). Nucleation is said to be instantaneous whenever the formation rate of a nucleus at a given site is expected to be at least 60 times greater than the expected rate of coverage of the site by growth only. Nucleation is said to be progressive when the expected coverage of a site by growth is at least 20 times greater than the coverage of the same site by the act of nucleation. After being formed, calcium orthophosphate nuclei can grow in one, two or three dimensions resulting in different shapes of the deposits like needles, disks or hemispheres depending on deposit/substrate binding energy and their crystallographic misfit. In the electrochemical deposition of HA from aqueous electrolytes, during the first approximately 12 min, the nucleation is instantaneous and is accompanied by a two-dimensional growth. Subsequently, the nucleation becomes progressive and is accompanied by a three-dimensional growth (Eliaz and Eliyahu 2007).

In general, calcium orthophosphate coatings, layers or films obtained by the electrochemical method have a uniform structure since they are formed gradually through a nucleation and growth process at relatively low temperatures (de Groot et al. 1998). Such coatings might be porous (Duan et al.2003). Interestingly, that in order to produce apatite coatings, non-apatitic calcium orthophosphates might be electrochemically deposited, followed by additional treatments (Redepenning et al. 1996; Han et al. 1999b, 2001; Kumar et al. 1999; Silva et al. 2001). Subsequently, the deposited calcium orthophosphate coatings, layers or films might be heat treated in water steam at 125°C (Shirkhanzadeh 1993) and/or then calcined at temperatures up to 800°C to densify and improve its bonding to the substrates.

Sol-gel deposition

By definition, a sol is a two-phase suspension of colloidal particles in a liquid, while gels are regarded as composites because they consist of a solid skeleton or network that encloses a liquid phase or an excess of the solvent. Therefore, the sol-gel process, as the name implies, is a wet-chemical technique that involves transition from a liquid 'sol' into a solid 'gel' phase. Colloidal particles can be in the approximate size range of 1 to 1,000 nm; hence, gravitational forces on these particles are negligible, and interactions are dominated by both short-range forces and surface charges. To prepare sols, usually, inorganic metal salts and/or organometallic compounds such as metal alkoxides are used as precursors. Sols are formed after a series of hydrolysis and condensation reactions of the precursors. Then, the sol particles condense into a continuous liquid gel phase. Besides, a sol might be prepared by dispersion of colloidal particles in a liquid, followed by destabilization of the sol to produce a particulate gel. With further drying and heat treatment, the gel is then converted into dense ceramic or glass materials (Morris 2011). The deposited gels create coatings, films and layers. Sol-gel coatings, films and layers are usually produced using spin or dip (Liu et al. 2002a) coating techniques (see below).

According to this technique, calcium orthophosphate coatings, layers or films are prepared by dipping the sample in calcium (usually, nitrate salt) and phosphorus (usually, alkyl phosphates) gels for an appropriate time at low reaction temperatures. As-formed coatings, layers or films are porous, less dense and have poor adhesion to the substrate. To improve their properties, the samples are annealed at temperatures of 400°C to 1,000°C (Gross et al. 1998c; Haddow et al. 1999; Montenero et al. 2000; Tkalcec et al. 2001; Liu et al. 2002b; Metikoš-Hukovi et al. 2003; Kim et al. 2004a; Gan and Pilliar 2004; Zhang et al. 2006; Stoica et al. 2008). Depending upon the temperature, different calcium orthophosphate compounds are obtained. The resulting coatings, layers or films can be extremely dense and adhere strongly to the underlying substrate (de Groot et al. 1998). Occasionally, in order to improve the bond strength between the coating and the substrate, an intervening layer of another compound might be applied prior the sol-gel deposition of a calcium orthophosphate (Kim et al. 2004b).

Biomimetic Deposition

Since biomimetics (synonyms: bionics, biomimicry) seeks to apply biological methods and systems found in nature, biomimetic deposition appears to be a method whereby a biologically active bone-like apatite layer is formed on a substrate surface by immersion in various simulating solutions, such as Hank's balanced salt solution (HBSS) or simulated body fluid (SBF) (Song et al. 2004; Habibovic et al. 2002; Oliveira et al. 2003; Hanawa and Ota 1992; Li et al. 1992; Leitão et al. 1995; Oliveira et al.1999; Wang et al. 2003). This method involves a heterogeneous nucleation and growth of bone-like calcium orthophosphate crystals on the surface of implants at physiological conditions (temperatures 25°C or 37°C and solution pH within 6 to 8) for several days or even weeks. However, since all simulating solutions, such as HBSS and SBF (their chemical composition might be found in literature) contain a number of various ions, ion-substituted calcium orthophosphates might be deposited only. The thickness of such calcium orthophosphate coatings, layers or films varies within several microns (Table 3), while, according to the X-ray diffraction measurements, the majority of the biomimetic precipitates appear to be either amorphous or poorly crystalline (de Groot et al. 1998). A typical example of a biomimetically deposited calcium orthophosphate coating is shown in Figure 7 (Layrolle2011).

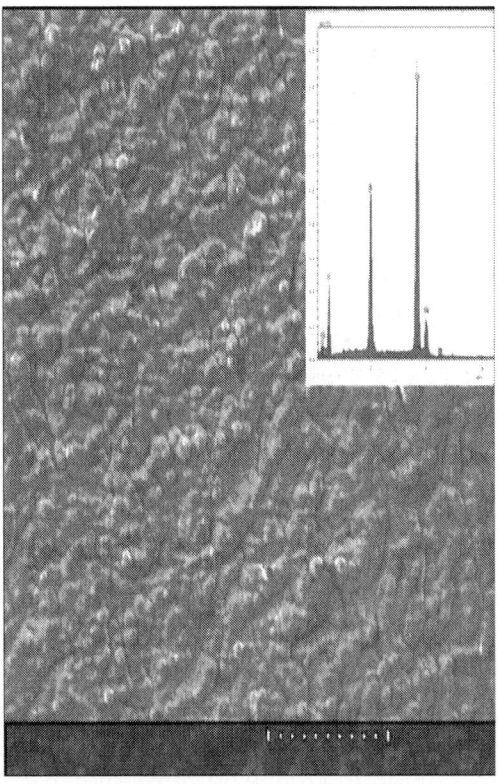

Figure 7: A scanning electron microscopy of a typical biomimetically deposited carbonated apatite coating. Inset: an EDX spectrum of the coating. Bar is 200 μm. Reprinted from Layrolle (2011) with permission.

The mechanism of bone-like apatite formation on an oxidized surface of titanium was investigated in details (Takadama et al. 2001; Uchida et al. 2003). Briefly, it looks as follows: First, a layer of amorphous sodium titanate is formed on the Ti surface after alkali pretreatment. Then, immediately after immersion into SBF, the sodium titanate exchanged Na^+ ions for H_3O^+ ions in the fluid to form Ti-OH groups on its surface. Later, the Ti-OH groups incorporated calcium ions from the SBF to form a layer of amorphous calcium titanate. After longer soaking times, the amorphous calcium titanate incorporated orthophosphate ions from the SBF to form ACP coatings with a Ca/P atomic ratio of approximately 1.4. Thereafter, ACP converted into bone-like ion-substituted CDHA with a Ca/P ratio of approximately 1.65, which was close to the value of bone mineral (Takadama et al. 2001). In the next study, the authors specified that, after exchanging Na^+ ions for H_3O^+ ions, various types of titania gels might be formed but only those with the anatase or rutile structure induced apatite formation (Uchida et al. 2003). Further specific details on this topic are available in literature (Kokubo and Yamaguchi 2011).

Since biomimetic deposition of calcium orthophosphate coatings, layers or films is a slow process, ways were sought to make it faster. Using condensed versions of the simulating solutions is the most popular option. For example, time for apatite induction in the 1.5-fold SBF was significantly shortened compared to that in the standard SBF. Therefore, the concentration of SBF was increased further, namely, 2-fold (Sun and Wang 2010; Miyaji et al. 1999; Kim et al. 2000), 5-fold (Barrere et al.2002a, 2002b, 2004) and even 10-fold (Tas and Bhaduri 2004) SBF solutions were used to accelerate precipitation and increase the amount of precipitates. However, whenever possible, this should be avoided because the application of condensed solutions of SBF leads to changes in the chemical composition of the precipitates; namely, the concentration of carbonates increases, while the concentration of orthophosphates decreases (Dorozhkina and Dorozhkin 2003).

The nucleation and growth of calcium orthophosphate coatings deposited on Ti6Al4V substrates from 5-fold SBF were investigated in details by both atomic force and environmental scanning electron microscopes (Barrere et al. 2004). Scattered calcium orthophosphate deposits of approximately 15 nm in diameter were found to appear after only 10 min of immersion in 5-fold SBF. Then, they grew up to 60 to 100 nm after approximately 4 h. With increasing immersion time, the packing of calcium orthophosphate deposits with size of tens of nanometers in diameter formed larger globules and then continuous calcium orthophosphate coatings on Ti6Al4V substrates. The coatings were composed of nano-sized deposits. A direct contact between calcium orthophosphates and the Ti6Al4V surface was observed (Barrere et al. 2004). A stable solution containing high concentrations of calcium and orthophosphate ions was prepared in another study (Li et al. 2002b). This solution became supersaturated after $NaHCO_3$ was added. A uniform coating of approximately 40-μm thickness was obtained on the substrate after immersion for 24 h. The coatings contained adjustable composition from CDHA to DCPD (Li et al. 2002b).

Simplification of the ionic composition of the standard simulating solutions is still another option to increase deposition kinetics (Bigi et al. 2005a; Li 2003). For example, a fast (a few hours instead of 14 days with SBF) biomimetic deposition of CDHA coatings on Ti6Al4V substrates was obtained using a slightly supersaturated Ca/P solution with an ionic composition simpler than that of SBF. Thin film XRD indicated that the deposits obtained after approximately 3 h were poorly crystalline CDHA, and their content increased on increasing the soaking time up to 3 days (Bigi et al. 2005a). However, since adhesion of this coating to the substrate was not indicated, it is doubtful whether this coating had sufficient strength to resist dissolution inside the body.

Dip Coating

Dip coating is a popular way of creating coatings, films and layers for various purposes. It consists of several successive steps. A substrate is immersed into either a solution or a suspension of the coating material (in our case, calcium orthophosphate) at a constant speed. A wet coating, film or layer is deposited by itself on the substrate while it is pulled up. Usually, withdrawing is carried out at a constant speed to avoid any jitters. The speed determines the thickness (the faster, the thinner). Excess solution or suspension is drained from the surface. A solvent evaporates from the solution or suspension, forming a denser coating, film or layer. For volatile solvents, such as alcohols, evaporation starts already during the deposition and drainage steps. After being dried and sintered, a solid surface is achieved (Brinker et al. 1991). By means of dipping, uniform coatings, films and layers of calcium orthophosphates have been applied onto various substrates (Li et al. 1996; Weng and Baptisa 1998; Jiang and Shi 1998; Choi et al. 2003; Bini et al. 2009).

There are two mechanisms which govern the formation of the surface coatings, films or layers during dip coating. The first mechanism is known as liquid entrainment. It occurs when a specimen is withdrawn from slurry faster than it can drain from the surface, leaving a thin film (Pontin et al.2005). The second mechanism is slip casting, in which the capillary suction caused by a substrate drives ceramic particles to concentrate at the substrate-suspension boundary, and a wet layer is formed (Gu and Meng 1999). The withdrawal velocity and the suspension properties (volume fraction of solids, viscosity) have influence on the liquid entrainment mechanism, while the surface microstructure of the substrate (porosity and pore diameter) together with the suspension properties have influence on the slip casting mechanism. By modifying these parameters, layers as thin as 2 µm and as thick as 0.5 mm might be formed (Pontin et al. 2005; Gu and Meng 1999).

Physical Vapor Deposition Techniques

In general, all types of physical vapor deposition (or sputtering) techniques for producing coatings, films and layers can be broadly classified into two main groups: (1) those involving thermal evaporation techniques, where a material is heated until its vapor pressure becomes greater than the ambient pressure, and (2) those involving ionic sputtering methods, where a highly energetic beam of ions and/ or electrons strikes a solid target and knocks atoms off from the surface (Narayanan et al.2010; Paital and Dahotre 2009a). Usually, physical vapor deposition occurs in vacuum; however, it might be performed in presence of some gasses. The target is the source material (in our case, a calcium orthophosphate). Substrates are placed into a chamber, and they are pumped down to a prescribed pressure. Sputtering is driven by momentum exchange between the ions and atoms in the materials due to collisions. Afterwards, the dislodged atoms or molecules are deposited on a substrate which is also placed into the same vacuum chamber. An important advantage of the sputter deposition is that even materials with very high melting points are easily sputtered. For the efficient momentum transfer, the atomic weight of the sputtering gas should be close to the atomic weight of the target, so neon or argon is preferable for sputtering of light elements, while krypton or xenon is used for heavy elements (Cuerno and Barabási 1995). However, for deposition of calcium orthophosphates, oxygen might be used as well. It has a number of features, and a better stoichiometry with respect to HA of the deposited coatings, films and layers is one of them (van Dijk et al. 1997).

To sputter calcium orthophosphates, several types of the physical vapor deposition techniques are used, such as ion beam (Stevenson et al. 1989; Barthell et al. 1989; Ong et al. 1991, 1992, 1994; Yoshinari et al. 1994; Cui et al. 1997; Kim et al. 1998; Luo et al. 1999; Choi et al. 2000; Wang et al.2001; Hamdi and Ide-Ektessabi 2003; Lee et al. 2003; Fujihara et al. 2004; Lee et al. 2005b; Rabiei et al. 2006; Lee et al. 2007a; Blalock et al. 2007), radio-frequency (RF) magnetron (Cooley et al.1992; Yamashita et al. 1994; Jansen et al. 1993; Wolke et al. 1994; van Dijk et al. 1995, 1996; Wolke et al. 1998, 2003; Nelea et al. 2003, 2004; Feddes et al. 2003a, 2003b; Ding 2003; Yamaguchi et al. 2006; Wan et al. 2007; Ozeki et al. 2007; Ueda et al. 2007; Snyders et al. 2008; Ievlev et al. 2008; Toque et al. 2009), pulsed laser (Nelea et al. 2000, 2002, 2004; Cotell et al. 1992, Cotell 1993; Torrisi and Setola 1993; Cotell 1993; Singh et al. 1994; Wang et al. 1997; Hontsu et al.1997; Fernández-Pradas et al. 1998, 1999, 2001; Mayor et al. 1998; Arias et al. 1998; Craciun et al.1999; Fernandez-Pradas et al. 1999; Zeng et al. 2000; Cleries et al. 2000a; Nelea et al. 2000; Zeng and Lacefield 2000; Fernandez-Pradas et al. 2001; Nelea et al. 2002; Socol et al. 2004; Kim et al.2005a; Bigi et al. 2005b; Koch et al. 2007; Kim et al. 2007a; Paital and Dahotre 2008; Paital et al.2009; Dinda et al. 2009; Tri and Chua 2009; Sygnatowicz and Tiwari 2009), diode, direct current and reactive sputtering or deposition (Massaro et al. 2001). The physical and aggregate states of the calcium orthophosphate source might influence the deposition kinetics. For example, the deposition rate of HA was found to be much higher in a solid plate target than in a powder lump target, owing to the difference of apparent density approximately 75% and approximately 18%, respectively (Wan et al. 2007).

Depending on the type of sputtering system and parameters used for the deposition, the structure and chemical composition of the deposited coatings, layers and films may be quite different from those of the initial material used for sputtering. For example, differences in

Ca/P ratios between the initial calcium orthophosphates and that in the sputtered coatings were suggested to be attributed to the preferential sputtering of calcium, probably due to a possibility of orthophosphate ions being pumped away before they are deposited at the substrate (Zalm 1989). It was also suggested that orthophosphate ions might be weakly bound to the growing coatings, layers and films, and therefore, they are sputtered away by incoming ions or electrons (van Dijk et al. 1996). Nevertheless, all sputtering techniques have the advantage of depositing thin coatings, films and layers with strong adhesion and compact microstructure.

Ion Beam Assisted Deposition

Ion beam assisted deposition (IBAD) is a vacuum technique in which ions of a material to be deposited (in our case, calcium orthophosphates) are generated by collisions with electrons. Then, the detached ions are accelerated by an electric field emanating from a grid toward a target. As the ions leave the source, they are neutralized by electrons from the second external filament and form neutral atoms. A pressure gradient between the ion source and a sample chamber is generated by placing a gas inlet at the source and shooting through a tube into the sample chamber (Ali et al.2010). Therefore, a typical deposition system consists of two main parts: electron or ion beam bombarding and vaporizing a calcium orthophosphate bulk target to produce an elemental cloud towards the surface of a substrate and a source for simultaneous irradiation of a substrate with highly energetic inert (e.g., Ar^+) or reactive (e.g., O_2^+) gas ions to assist the deposition. Both single and dual ion beam assisted deposition systems are available. Good illustrations of both systems are presented in literature (Narayanan et al. 2010; Paital and Dahotre 2009a; Surmenev 2012).

In this approach, firstly thin (a few hundred atomic layers thick) and amorphous calcium orthophosphate coatings, layers or films are usually deposited. Then, an ion implantation technique, with ions such as argon, nitrogen and oxygen, is used to make them crystalline (Stevenson et al.1989; Barthell et al. 1989; Ong et al. 1991, 1992, 1994; Yoshinari et al. 1994; Cui et al. 1997; Kim et al. 1998; Luo et al. 1999; Choi et al. 2000; Wang et al. 2001; Hamdi and Ide-Ektessabi 2003; Lee et al. 2003, 2007a; Fujihara et al. 2004; Lee et al. 2005b; Rabiei et al. 2006; Blalock et al. 2007). A high bond strength associated with this deposition technique appears to be a consequence of an atomic intermixing interfacial layer, which can be up to a few microns thick. Studies revealed alterations in the chemical composition of the ion beam deposited coatings, layers or films. For example, calcium orthophosphate films were synthesized on silicon wafers by electron beam evaporation of -TCP both with and without simultaneous Ar ion beam bombardments (Lee et al.2003). It was observed that a simultaneous bombardment with Ar ion beam had a significant effect on both the morphology (Figure 8) and composition of the films, namely, films formed without Ar ion beam bombardment were found to have a Ca/P ratio of approximately 0.76 and reacted immediately with the moisture in the air as soon as it is removed from the chamber. In contrast, the films formed with Ar ion beam bombardment had a Ca/P ratio of approximately 0.80 with smooth and featureless surface morphology (Lee et al. 2003).

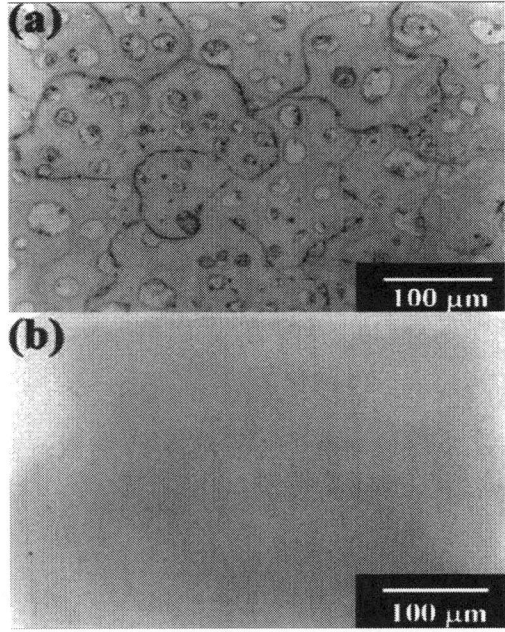

Figure 8: Optical micrographs of a calcium orthophosphate layer deposited on a Si wafer. (a) Without ion beam bombardments and (b) with Ar ion beam bombardments (120 V, 0.8 A). Reprinted from Lee et al. (2003) with permission.

In another study, calcium orthophosphate layers on silicon substrates were prepared by using ion beam assisted simultaneous vapor deposition. The method comprised of an electron beam heater and a resistance heater vaporizing CaO and P_2O_5, respectively, while an argon ion beam was focused onto substrates to assist the deposition (Hamdi and Ide-Ektessabi 2003). All deposited layers appeared to be amorphous, regardless of the current density level of the ion beam. Therefore, a post-heat treatment was applied to crystallize the layers. The effects of ion beam current density on the phase composition of the crystallized calcium orthophosphates are shown in Figure 9. The Ca/P ratio was found to increase with increasing ion beam current density presumably due to the high sputtering rate of P_2O_5 compared to that of CaO from the layer being coated. As seen in Figure 9, biphasic (HA + TCP) formulations were found when the ion beam was either not used or used at current density of 180 mA/cm^2, while at ion beam current density of 260 mA/cm^2, only HA peaks were observed (Hamdi and Ide-Ektessabi 2003). In still another study, the X-ray photoelectron spectroscopy analysis of the deposited calcium orthophosphate coatings on titanium revealed several distinct zones:

- the ambient-exposed surface exhibited elevated concentrations of carbon due to atmospheric contamination;
- the bulk zone contained relatively constant concentrations of calcium, oxygen, phosphorus and fluorine, indicating the chemistry for calcium fluoride and FA formation;
- While the underlaying zone exhibited elevated titanium and oxygen photoelectron peaks, suggesting the coexistence of calcium orthophosphates within titanium oxides. Furthermore, the substrate was shown to be identical to the passivated titanium surface prior to deposition (Ong et al. 1991). A similar zone structure was also discovered by other researchers (Wang et al. 2001). In addition, a cross-section of functionally graded thin HA coatings on silicon substrate obtained by a dual ion beam assisted deposition and simultaneous heat treatment was investigated and the microstructural analysis of the coatings revealed a gradual decrease of the grain size and crystallinity towards the surface, leading to nano-scale grains and eventually amorphous layer at the surface (Rabiei et al.2006).

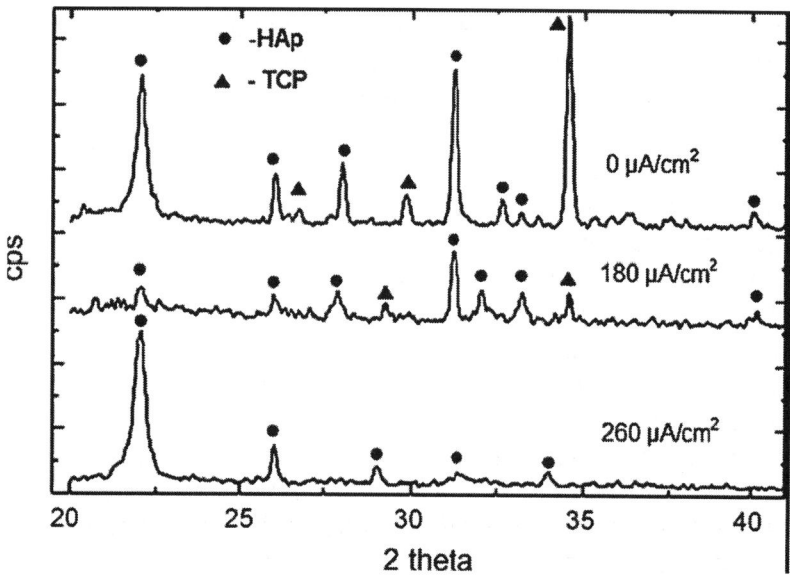

Figure 9: XRD patterns of fully crystallized (after a heattreatment at 1200°C) calcium orthophosphate coatings. Sputtered at three different values of ion beam current density. Reprinted from Hamdi and Ide-Ektessabi (2003) with permission. HAp, hydroxyapatite; TCP, tricalcium phosphate.

Choi et al. (2000) deposited HA films on Ti-6Al-4V alloy by electron beam vaporization of pure HA target and simultaneous bombardment using a focused Ar ion beam on the metal substrate to assist deposition. The effect of Ar ion beam current on the bond strength and dissolution of the coating in a physiological solution was studied. The bond strength between the coating and the substrate increased with increasing current, whereas the dissolution rate in physiological solution decreased remarkably (Choi et al. 2000). Further details on this technique are available in the aforementioned references.

Pulsed Laser Deposition or Laser Ablation Deposition

Shortly after the discovery of a laser in the end of 1950s (Amy and Storb 1955), researchers began focusing their beams at materials to observe the interaction. PLD or laser ablation deposition technique for producing thin films became increasingly popular in 1970s due to the advent of lasers delivering nanosecond pulses (Singh and Narayan 1990). In this technique, a high power pulsed laser beam is focused inside a vacuum chamber to strike a target of the material (in our case, calcium orthophosphates) resulting in a gaseous cloud of various atoms, ions, molecules, molecular clusters and, in some cases, droplets and target fragments, due to a thermal decomposition of the target (Dinda et al. 2009). For sufficiently high laser energy density, each laser pulse vaporizes or ablates a small amount of the material, creating a plasma plume. The ablated material is ejected from the target in a highly forward-directed plume. The ablation plume provides the material flux, which then is deposited on a substrate. This process can occur in both ultra-high vacuum and presence of a background gas, such as oxygen, which is commonly used when depositing oxides to fully oxygenate the deposits. Argon (Bao et al. 2006) and water vapor (Fernandez-Pradas et al. 2000) can be used as well. The thorough investigation of a plasma plume expansion process during an ArF laser ablation of HA is well described elsewhere (Jedynski et al. 2008). The experimental setup of a PLD technique is available in literature (Narayanan et al. 2010; Paital and Dahotre 2009a; Surmenev 2012); it essentially consists of a laser source, an ultrahigh vacuum deposition chamber equipped with a rotating target and a fixed substrate holder plus pumping systems. Mostly, the substrates are attached to the surface parallel to the target surface at a target-to-substrate distance of 2 to 10 cm. Usually, for ablation, ultraviolet excimer lasers with pulses of approximately 10-ns duration and power densities in the order of 10 to $500\,MW/cm^2$ are required. A two-laser beam technique (so-called, laser-assisted laser ablation method) is used as well (Katayama et al. 2009). In this technique, one laser beam from KrF laser, the ablation laser, is used for ablation of a HA target. The other beam from ArF laser, the assist laser, is used to irradiate a Ti substrate surface during formation of the HA coating. The assist laser plays an important role in the formation of a crystalline HA coating and improves the strength of adhesion to the Ti substrate (Katayama et al. 2009). Further details on the PLD technique might be found in literature (Willmott and Huber 2000).

A PLD process is used for forming thin (0.05 to 5 μm) calcium orthophosphate coatings, layers or films on various substrates (Nelea et al. 2000, 2002, 2004; Cotell et al. 1992, 1993; Torrisi and Setola 1993; Singh et al. 1994; Wang et al. 1997; Hontsu et al. 1997;

Fernández-Pradas et al. 1998,1999, 2001; Mayor et al. 1998; Arias et al. 1998; Craciun et al. 1999; Fernandez-Pradas et al. 1999; Zeng et al. 2000; Cleries et al. 2000a; Zeng and Lacefield 2000; Socol et al. 2004; Kim et al. 2005a,2007a; Bigi et al. 2005b; Koch et al. 2007; Paital and Dahotre 2008; Paital et al. 2009; Dinda et al.2009; Tri and Chua 2009; Sygnatowicz and Tiwari 2009). The process involves ablation of a calcium orthophosphate (usually, HA) target using a pulsed (usually, pulses of 30 ns and 120 mJ at a repetition of 10 Hz) KrF excimer laser beam (= 248 nm) in 0.3 Torr/H_2O atmosphere and deposition of the ejected HA material on a heated (400°C to 800°C) substrate. The deposition rate of PLD is about 0.02 to 0.05 nm per laser shot (de Groot et al. 1998). An investigation into the effects of high laser fluence (between 2.4 J/cm^2 and 29.2 J/cm^2) on the properties of calcium orthophosphate films was performed (Tri and Chua 2009). The films deposited at 2.4 J/cm^2 were found to be partially amorphous and had rough surfaces with many droplets, while higher laser fluences showed a higher level of crytallinity and lower surface roughness. Furthermore, higher laser fluences also decreased the ratio Ca/P of as-deposited films and, probably, increased their density (Tri and Chua 2009). The substrate heating is necessary to ensure the formation of a highly crystalline and phase pure coatings, films and layers. Besides, the substrate temperature could be varied to provide deposits with the desired fine texture and roughness, depending on their application (Saju et al. 2009; Rau et al. 2010).

Typically, during the deposition, a target should be rotated to achieve a stable ablation rate. As PLD is usually carried out at high substrate temperatures, a thin oxide layer might be formed on the substrate surface prior to the deposition of calcium orthophosphates and, thereby, it influences its adherence to the substrate (Nelea et al. 2000). The deposited coatings, layers and films frequently consist of several calcium orthophosphates (often with admixtures of other substances, such as CaO, calcium pyrophosphates, etc.) and might contain both amorphous and crystalline phases (Cleries et al. 2000a; Koch et al. 1990). Biphasic formulations, such as HA + TTCP (Kim et al. 2005b), might be deposited as well. Interestingly, TTCP in the coatings was not formed by partial conversion of previously deposited HA. Instead, it was produced by nucleation and growth of TTCP itself from the ablation products of the HA target or by accretion of TTCP grains formed during ablation (Kim et al.2005b). Furthermore, the PLD-deposited coatings, films and layers might consist of calcium orthophosphates with different morphologies (e.g., granular and columnar), which have different resistance values to delamination (Cleries et al. 2000a). More to the point, various types of oriented textures might be created as well (Kim et al. 2005a, 2007a, 2010). A modification known as 'transmission laser coating' has been introduced (Cheng and Ye 2010).

Further details on the PLD technique might be found in literature (Surmenev 2012; Koch et al. 2007).

Magnetron Sputtering

A magnetron is a high-powered vacuum tube that generates microwaves using the interactions of a stream of electrons with a magnetic field. Magnetron sputtering technique has been emerged in the mid of 1960s (Gill and Kay 1965) and is considered as a high-rate vacuum coating technique for depositing metals, alloys and compounds onto a wide range of materials with thickness up to approximately 5 µm (Table 3). A sputtering system consists of an evacuated chamber, a wave generator, a magnetron, a cooling system, as well as it contains a target and a substrate. It works on the principle of applying a specially shaped magnetic field to a sputtering target. Once the substrate is placed into the vacuum chamber, air is removed, and the target material (in our case, calcium orthophosphates) is released into the chamber in the form of a gas. Powerful magnets ionize particles of the target material. Then, the negatively charged target material lines up on the substrate to form deposits (Kelly and Arnell 2000). In principle, magnetron sputtering can be done in either DC (direct current) or RF modes; however, since the DC mode might be done with conducting materials only, the RF mode is solely used to deposit calcium orthophosphates. Typically, RF magnetron sputtering employs a sinusoidal wave generator operating at 13.56, 5.28 or 1.78 MHz. The parameters that directly affect the quality and integrity of calcium orthophosphate coatings, films and layers include discharge power, gas flow rate, working pressure, substrate temperature, deposition time, post-heat treatment and negative substrate bias (Surmenev 2012). For example, the deposition rate was found to increase with increasing argon gas pressure up to 2 Pa but decreased significantly as the pressure increased up to 5 Pa, while the Ca/P ratios of as-deposited coatings decreased significantly at the higher argon gas pressures (Boyd et al. 2007). A good schematic setup of a magnetron sputtering system is available in literature (Narayanan et al. 2010; Paital and Dahotre 2009a; Surmenev 2012).

For the first time, RF magnetron sputtering was used to prepare calcium orthophosphate coatings in 1992 (Cooley et al. 1992). Since then, it has become a convenient method for deposition of biocompatible ceramic coatings, layers and films on various substrates (Yamashita et al. 1994; Jansen et al. 1993; Wolke et al. 1994, 1998, 2003; van Dijk et al. 1995, 1996; Nelea et al. 2003,2004; Feddes et al. 2003a, 2003b; Ding 2003; Yamaguchi et al. 2006; Wan et al. 2007; Ozeki et al.2007; Ueda et al. 2007; Snyders et al. 2008; Ievlev et al. 2008; Toque et al. 2009). The advantages of magnetron sputtering over other sputtering processes include a high deposition rate, an excellent adhesiveness and an ability to coat implants with difficult surface geometries (Table 3). Still, several issues, such as the endurance and the Ca/P ratio, have to be solved before magnetron sputtering can be applied to deposit, on a routine basis, pure and crystalline calcium orthophosphates on implant surfaces. For example, both microstructure and mechanical properties of HA thin films, grown on Ti-5Al-2.5Fe alloys by RF magnetron sputtering, were investigated (Nelea et al. 2003). The deposition was performed from pure HA target in low pressure Ar or Ar-O_2 mixtures at substrate temperatures ranging from 70°C to 550°C. Smooth (an average roughness of approximately 50 nm) and uniform calcium orthophosphate films were fabricated. It was observed that the films grown at the substrate temperatures below approximately 300°C were prevalently amorphous (ACP) and contained a small amount of crystalline phases. On the contrary, the films obtained at a substrate temperature of 550°C or the films grown at room temperature followed by annealing at 550°C consisted of HA (Nelea et al. 2003).

The chemical composition of the deposited calcium orthophosphate coatings, films and layers might be modified by varying the RF sputtering power density (Snyders et al. 2008), namely, when the power density was increased by 240%, the Ca/P ratio increased from approximately 1.51 to approximately 1.82. X-ray diffraction indicated the phase pure HA except for the samples prepared at the highest power density values, in which the presence of CaO and TCP was also detected. Interestingly, deviations from the stoichiometric HA resulted in reduction of the elastic modulus, namely, for Ca/P approximately 1.51, the elastic modulus dropped by approximately 15%, which was attributed to Ca vacancies in the lattice, while for Ca/P approximately 1.82, the average elastic modulus decreases by approximately 10% due to formation of additional phases (Snyders et al.2008).

Various types of calcium phosphates were magnetron sputtered from TTCP, HA, -TCP, -calcium pyrophosphate (CPP) and -calcium metaphosphate(CMP)powdertargets(Ozekietal.2007).Thecomposition of the deposited films was changed depending on the target materials, whiletheCa/Pmolarratiosofthefilmsvariedfrom0.74to2.54,increasing with the Ca/P molar ratio of the target. Interestingly, the deposition rate of the aforementioned calcium phosphates was established as the following: TTCP ≈ β-CMP > β-TCP > β-CPP ≈ HA, which correlated well to the solubility order: TTCP ≈ β-CMP > β-TCP > β-CPP ≈ HA (Ozeki et al. 2007).

RF magnetron sputtering might be combined with other deposition techniques. For example, plasma-assisted RF magnetron co-sputtering deposition method was used to deposit calcium orthophosphates on Ti6Al4V orthopedic alloy (Xu et al. 2005; Long et al. 2007). Further details on the magnetron sputtering technique are available in excellent reviews (Surmenev 2012; Shi et al. 2008).

Other Deposition Techniques: Miscellaneous

Prior describing the below mentioned deposition techniques, one should note that they are rare and are mentioned in just a few research papers. Therefore, the detailed description is not always possible.

Hot Isostatic Pressing

Hot isostatic pressing (HIP) is a manufacturing process used to reduce porosity and increase the density of many types of materials. The HIP process subjects a component to both elevated temperature and isostatic gas pressure in a high-pressure containment vessel. To deposit coatings, initially, solid cores are covered by a calcium orthophosphate (usually HA) powder. Both organic binders and some other additives are used to simplify deposition. A furnace is constructed within the high-pressure vessel, and the coated samples are placed inside to be pressed. Then, the specimens are heated at temperatures within 700°C to 1,200°C and pressed at pressures within 20 to 100 MPa. The obtained coatings, films and layers are usually thick (0.2 to 2.0 mm) and dense (Bocanegra-Bernal 2004).

The HIP technique was used to manufacture calcium orthophosphate coatings, films and layers on various materials (Lacefield 1988; Herø et al. 1994; Wie et al. 1998; Kameyama 1999). For example, HA granules (32 to 38 μm in diameter) were implanted into a substrate of superplastic titanium alloy. First, the HA granules were spread over this surface and, then, hot pressed at 750°C and 17 MPa for 1 h with a plunger to implant them into the substrate. After 10 min of implantation, the implantation ratio was approximately 20%, and some granules were not on the substrate. After 60 min of implantation, the implantation ratio was 100%, but the upper areas of granules were exposed (Kameyama 1999).

A variation in the HIP technique was proposed in which thin HA coatings were prepared with a curved surface at low temperatures (Onoki and Hashida 2006). The method used double-layered capsules in order to create suitable hydrothermal conditions; the inner capsule encapsulated the coating materials and a Ti substrate, while the outer capsule was subjected to isostatic pressing under the hydrothermal conditions. It was demonstrated that a HA layer of approximately 50 μm thick could be deposited on Ti cylindrical rods at 135°C under the confining pressure of 40 MPa. The deposited HA layer had a porous microstructure with the relative density of approximately 60%. According to the results of pullout tests, the shear strength was in the range of 4.0 to 5.5 MPa. These results also revealed that a crack propagation occurred within the HA coating layer but not along the HA/Ti interface. This observation suggests that the fracture property of the HA/Ti interface was higher than that of the HA ceramics only. Thus, hydrothermal HIP technique appears to be a useful method for producing bioactive HA ceramic coatings on curved prostheses surfaces (Onoki and Hashida 2006). However, the majority of the calcium orthophosphate coatings, layers or films produced by HIP technique are contaminated with metal and SiO_2 particles due to the use of a glass encapsulating tubes (de Groot et al. 1998).

Frit Enameling

Frit is a ceramic composition that was fused in a special fusing oven, quenched to form a glass and granulated. According to this technique, a metal bar was dipped into a HA slurry, dried and sintered at 1,100°C to 1,200°C for a minimum of 3 h in a protective Ar atmosphere (Lacefield 1988). Coatings created in this technique show very low interfacial shear strength (approximately 0.22 MPa). Probably, the furnace atmosphere used for sintering was inadequate, which resulted in excessive substrate oxide layer thickness and poor bonding (de Groot et al. 1998).

Aerosol-Gel

An aerosol is a colloid suspension of fine solid particles or liquid droplets in a gas. Examples are clouds and air pollution, such as smog and smoke. While gels are regarded as composites because they consist of a solid skeleton or network that encloses a liquid phase or an excess of the solvent. Therefore, the aerosol-gel process, as the name implies, is a gas-chemical technique that involves transition from a gaseous 'aerosol' into a solid 'gel' phase.

The aerosol-gel technique was also applied to produce highly porous calcium orthophosphate coatings on various materials (Manso et al. 2002a; 2002b; Manso-Silván et al. 2003). Calcium nitrate and triethylphosphate diluted in ethanol were used as precursor solutions. After production of a steady state aerosol, the microdroplets were conducted into a deposition chamber by an air flux. After being deposited, the coatings were sintered at temperatures within 500°C to 1,000°C. The composition, structure and morphology of the final coatings were found to fit highly porous polycrystalline HA. The adhesive strength, measured by means of indentation techniques, was found to be in the order of 100 MPa, which was significantly higher than the values obtained by sol-gel deposition technique (Manso et al. 2002a; 2002b; Manso-Silván et al. 2003).

Micro-Arc Oxidation

Micro-arc oxidation (MAO), also called plasma electrolytic oxidation, anodic spark deposition, or micro-arc discharge oxidation, is a plasma-chemical and electrochemical process. The process combines electrochemical oxidation with a high-voltage spark treatment in an aqueous electrolytic bath, which also contains modifying elements in the form of dissolved salts (*e.g.*, silicates) to be incorporated into the resulting coatings. A schematic setup of a MAO system is available in literature (Narayanan et al. 2010; Paital and Dahotre 2009a).

By means of a MAO technique, calcium orthophosphate coatings, films and layers have been prepared on various metals (Song et al. 2004; Liu et al. 2005; Sun et al. 2007; Wei et al. 2007; Han et al.2008). For example, MAO was performed on titanium in an electrolyte containing calcium glycerophosphate and calcium acetate using a direct current power supply. The MAO technique appeared to be suitable to form porous and rough ceramic coatings containing Ca and P. Then, the coatings were hydrothermally treated in aqueous solutions with pH within 7.0 to 11.0 (adjusted by adding NaOH) at 190°C for 10 h in an autoclave. This procedure converted undisclosed calcium orthophosphates into CDHA and/or HA crystals, while the amount of precipitated HA increased with solution pH increasing (Liu et al. 2005).

Direct Laser Melting

According to the direct laser melting technique, a starting precursor (calcium orthophosphate) powder is mixed thoroughly in a water-based organic solvent. Then, the suspension is sprayed onto the substrate surfaces to create a coating, which is then air dried to remove the moisture, followed by direct laser melting using either continuous wave or pulsed laser beams to produce strong bonds between the calcium orthophosphate coating and the substrate (Paital and Dahotre 2007, 2008,2009a, 2009b; Kurella and Dahotre 2006. A schematic setup of a direct laser melting system is available in literature (Paital and Dahotre 2009a).

Laser Cladding

Calcium orthophosphate coatings, films and layers were synthesized on various substrates by laser cladding (Lusquiños et al. 2001, 2003, 2005). Cheap precursors, such as mixed powders of calcium carbonate and DCPA/DCPD, can be used to prepare coatings (Wang et al. 2008b; Zheng et al. 2010; Lü et al. 2011a, 2011b; Lv et al. 2012). The reactions between $CaCO_3$ and DCPA/DCPD can produce high crystallized HA in the coatings, as well as TTCP, -TCP, -TCP, $Ca_2P_2O_7$, and CaO, if the Ca/P ratio is deviated from 1.67. Since the reactions between the powders produce gaseous by-products (CO_2 and water vapor), the prepared coatings were porous. Furthermore, when deposition was performed on metals (e.g., Ti), other admixtures, such as $CaTiO_3$, were formed. As the laser power increased, the amount of TTCP, HA and CaO in the coatings decreased gradually and, finally, only -TCP and $CaTiO_3$ remained. Nevertheless, the amount of HA could be increased greatly by heat treatment at 800°C for 5 h followed by furnace cooling, due to the total transformation of TTCP and -TCP to HA (Wang et al. 2008b; Zheng et al. 2010; Lü et al. 2011a, 2011b; Lv et al. 2012).

Detonation Gun Spraying

Detonation gun spraying is a high temperature and a high velocity technique which is thought to introduce a higher degree of melting to starting powder. This technique has also been explored to prepare calcium orthophosphate coatings on titanium alloys (Gledhill et al. 1999, 2001). In this process, a mixture of oxygen and acetylene is fed into a barrel together with a charge of the power. The gas is ignited, and detonation waves accelerate the powder up to about 750 m/s. The process produces a denser coating, which has a higher proportion of the amorphous phase with some evidence for the appearance of -TCP. A lower crystallinity and higher residual stress found in the detonation gun sprayed coatings resulted in a faster dissolution rate both *in vitro* and *in vivo* (Gledhill et al. 1999, 2001).

Cold Spraying Technique

In recent years, a new coating technology, known as cold spraying, has been developed (Gärtner et al. 2006). In this process, spraying particles (1 to 50 μm in size) experiencing both a little change in microstructure and a little oxidation or decomposition are accelerated by a supersonic jet of a compressed gas stream passing through a Laval type nozzle to a very high velocity (300 to 1,200 m/s). The deposition system consists of a gas pressure regulator, a gas heater, a powder feeder and a spray gun. In this technique, the deposited calcium orthophosphate particles are always in a solid state and at temperatures below their melting point. Thus, all phenomena occurring at high temperatures, such as thermal decomposition and phase transformations (see section Thermal spraying techniques above), are avoided. Deposition of calcium orthophosphate particles takes place through intensive plastic deformations (Zhang and Zhang 2011; Zhang et al. 2012a, 2012b). However, for successful bonding, the deposited particles have to exceed a critical velocity on impacts, which is dependent on the properties of the particular sprayed material (Gärtner et al. 2006).

Thermal Substrate Deposition

Thermal substrate deposition technique is based on the solubility differences at low and high temperatures, namely by heating a substrate in suitable saturated aqueous solutions, coatings, films and layers can be directly deposited onto the substrate. Various heating techniques of substrates have been proposed, namely conductive substrates, such as foil or wire, can be heated by electric current through them. Non-contact techniques, such as high frequency induction, can be used to heat materials with complex shapes. In either case, the immersed metallic sample can be heated up to 160°C in solutions, giving local supersaturations to perform crystallization.

Using this approach, calcium orthophosphate coatings, films and layers were obtained on titanium (Ziani-Cherif et al. 2002; Okido et al. 2002; Kuroda et al. 2002a, 2002b, 2003, 2005). An alternating current is passed through the metallic sample immersed in an aqueous solution containing calcium and phosphorus compounds. The deposition is usually performed for 10 to 30 min at solution pH 4 to 8. The type of precipitates varies, depending on the solution pH, temperature and ion concentrations, namely, precipitates of high quality, whose predominant component was HA (at pH > 6) or DCPA (at pH = 4), were obtained on titanium substrates by this technique. The content of HA in the deposits was found to increase with increasing temperature and heating time (Ziani-Cherif et al. 2002; Okido et al. 2002; Kuroda et al. 2002a, 2002b, 2003, 2005).

Matrix-Assisted Pulsed Laser Evaporation

Matrix assisted pulsed laser evaporation technique was developed as an alternative to PLD for delicate and accurate deposition of calcium orthophosphate films, coatings and layers combined with organic and/or biologic materials. The examples include deposition calcium orthophosphate-based biocomposites with sodium maleate (Negroiu et al. 2008), alendronate (Bigi et al. 2009) and silk fibroin (Miroiu et al. 2010). Thermally unstable calcium orthophosphates, such as OCP (Boanini et al.2012), might be deposited as well. This technique provides a more gentle mechanism for transferring different compounds, including large molecular weight species, and it is expected to ensure an improved stoichiometric transfer, a more accurate thickness control and a higher uniformity of the coatings.

Electrostatic Spray Deposition

Electrostatic spray deposition (ESD) is based on generation of an aerosol composed of organic solvents containing inorganic precursors under the influence of high voltages. According to this technique, spray droplets are generated by pumping a solution through a nozzle. Between the nozzle and substrate, a high voltage is applied. Consequently, droplets coming out the nozzle are dispersed into a spray, and this spray is deposited upon the substrate. When the solvent has evaporated, a coating is formed. The schematic setup of the ESD technique is available in literature (Leeuwenburgh et al. 2003, 2004, 2005a).

To perform ESD, a soluble calcium salt (nitrate or chloride) and phosphoric acid were dissolved in an alcohol. The obtained solutions were pumped, quickly mixed prior the nozzle and electrostatically sprayed onto a substrate, while the substrate itself might be heated to 300°C to 450°C (Leeuwenburgh et al. 2003, 2004, 2005a, 2005b, 2006a, 2006b). Besides, calcium orthophosphate powders might be suspended in alcohols, and the obtained suspensions are electrostatically sprayed (Lee et al. 2007b; Jiang et al. 2008; Iafisco et al. 2012). The chemical and morphological characteristics of the deposited calcium orthophosphate coatings, films and layers were found to be strongly dependent on both the composition of the precursor solutions (pH, absolute and relative precursor concentrations) and the deposition parameters, such as temperature, the nozzle-to-substrate distance, the liquid flow rate, as well as the geometry of the spraying nozzle. By varying these parameters, several phases and phase mixtures might be deposited by ESD technique: carbonate apatite, carbonated HA, -TCP, -TCP, DCPA, - and -calcium pyrophosphates, calcium metaphosphate, $CaCO_3$, CaO (Leeuwenburgh et al. 2005b, 2006a, 2006b). Since ESD might be performed at ambient temperatures, thermally unstable compounds could be deposited. As seen in Figure 10, the electrostatically sprayed calcium orthophosphate coatings, layers and films might be porous (Leeuwenburgh et al. 2006a, 2006b; Lee et al. 2007b; Jiang et al. 2008; Iafisco et al. 2012; Zhu et al. 2012). Nevertheless, after the deposition, the coated samples might be annealed at high temperatures. The annealing stage is necessary to aggregate and/or melt the deposited calcium orthophosphate particles and form highly dense and homogeneous coatings.

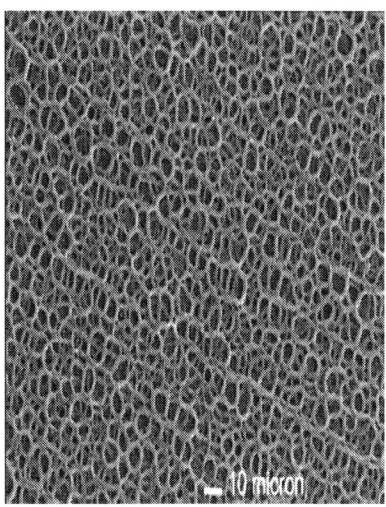

Figure 10: A scanning electron microscopy of an electrostatic spray deposited calcium orthophosphate coating. Characterized by a porous surface morphology. Reprinted from Leeuwenburgh et al. (2006a) with permission.

To conclude this part, combined techniques, such as sol-gel-assisted electrostatic spray deposition (Kim et al. 2007b) and electrostatic spray-assisted vapor deposition (Hou et al. 2007), have been developed as well. Further details on the ESD techniques are available in the aforementioned references.

Spin Coating

Spin coating is a procedure used to apply uniform thin films to flat substrates. It is rather similar to dip coating. The coating process consists of four stages: deposition, spin up, spin off and evaporation.

In this process, a sample is dipped in a solution or suspension and then withdrawn at a constant speed, usually with the help of a motor. Rotational draining and solvent evaporation result in the deposition of a coating, film or layer. A machine used for spin coating is called a spin coater, or simply spinner (Mennicke and Salditt 2002). Just a few publications on spin coating of calcium orthophosphates were published (You et al. 2005; Yuan et al. 2009; Carradò and Viart 2010).

Properties

Generally, for clinical applications, slowly or non-resorbable high crystalline coatings, films and layers have been recommended in order to retain the bonding strength with implants. However, this contradicts to the statement that the ideal interface between the implants and surrounding tissues should match the tissues being replaced. For example, in the case of HA, its crystallinity has been stated to be in the inverse proportion to its bioactivity (LeGeros 2008). Therefore, from the bioactivity point of view, calcium orthophosphate coatings, films and layers should be of low crystallinity and also contain various ionic substitutions, such as sodium, magnesium and carbonate. Thus, since one of the first steps in bonding involves dissolution of the coating surface, it might be suggested that coatings, films and layers prepared from less crystalline and/or more resorbable calcium orthophosphates would be more beneficial for early bone ingrowth than those prepared from high crystalline HA (Narayanan et al. 2010). However, soluble coatings, films and layers will weaken the bonding strength between them and substrates. In particular, a rapid dissolution of coatings, films and layers may loosen the bonding strength between the implant surface and the host bone. For example, a comparative study on the biological stability and osteoconductivity of HA coatings on Ti produced by pulsed laser deposition and plasma spraying was conducted. After 24 weeks of implantation, the plasma sprayed HA coatings showed considerable instability and reduction in thickness but no statistical difference to the uncoated Ti (the control), while the pulsed laser deposited ones remained almost intact but showed a significantly higher amount of bone apposition (Peraire et al. 2006). Thus, in that study the coating stability prevailed over its solubility. Furthermore, the excessive amount of the dissolved ions from the soluble coatings, films and layers may cause local inflammatory reactions.

Except for crystallinity and chemical composition, a number of other factors appear to influence the physical, chemical and mechanical properties of calcium orthophosphate coatings, films and layers. They include thickness (this will influence adhesion and fixation - the agreed optimum now seems to be within 50 to 100 μm), phase and chemical purity, fatigue resistance, porosity and adhesion (de Groot et al. 1998; Sun et al. 2001). Abrasion resistance might be important as well (Morks et al.2007).

Fatigue Properties

Several studies have already demonstrated that cyclic loading of the coated samples leads to fatigue failure. Further, it has been shown that a combination of an aqueous environment with stress can result in delamination or accelerated dissolution of calcium orthophosphate coatings, which can influence the long-term stability of the implants (Kummer and Jaffe 1992; Reis et al. 1994; Wolke et al. 1997). For example, calcium orthophosphate coatings were RF magnetron sputtered on Ti-6Al-4 V bars and, afterwards, some of them were annealed at 650°C to convert ACP into crystalline structure. Then, the coated samples were mechanically tested in either dry or wet (SBF solution) conditions (Wolke et al. 1997). The results of SEM demonstrated that, after cyclic loading conditions in air, the bars coated by crystalline calcium orthophosphates showed a partial coating loss. Furthermore, in wet conditions only the heat-treated sputter-coated bars appeared to be stable. On the other hand, the ACP coatings showed signs of delamination in more stressed regions only (Wolke et al. 1997). Thus, the fatigue properties of amorphous and crystalline calcium orthophosphate coatings, films and layers are different. Furthermore, the fatigue behavior shows substantial differences when tested in either dry or wet/conditions.

Thickness

Depending on the deposition technique, the thickness of the calcium orthophosphate coatings, films and layers varies from nanometric dimensions to several millimeters (Table 3), and this parameter appears to be very important, namely, if calcium orthophosphate coatings, films and layers are too thick, they are easy to break. Furthermore, the outer layers might tend to detach from the inner ones. On the contrary, if calcium orthophosphate coatings, films and layers are too thin, they are easy to dissolve because resorbability of HA, which is the second least soluble among calcium orthophosphates (Table 1), is about 15 to 30 μm per year under the physiological conditions (Gineste et al. 1999). To complicate the situation, the failure mechanisms for thinner and thicker coatings, films and layers appear to be different, namely the failure mode of thinner (50 μm) HA coatings on a Ti alloy was found to be conclusively at or near the coating/bone interface, while that of thicker (200 μm) HA coatings was found to be at the coating/bone interface, inside the HA lamellar splat layer, as well as at the coating/Ti alloy substrate interface (Wang et al. 1993; Yang et al. 1997). A similar conclusion was made in another study, in which the mechanical behavior of thin (0.1, 1 and 4 μm) calcium orthophosphate coatings was compared (Vercaigne et al. 2000a). Considering these points, commercial plasma-sprayed HA coatings, films and layers have thicknesses between 50 and 200 μm (Sun et al. 2001), though cells and tissues interact with only top surface, and thus, thickness of approximately 10 nm would be sufficient for cell activity.

Calcium orthophosphate coatings of various thickness ranging from 170 nm up to 1.5 μm were obtained depending upon the deposition times (Fernandez-Pradas et al. 2001). The coating morphology was found to be grain-like particles and droplets. During growth, the grain-like particles grew in size, partially masking the droplets, and a columnar structure was developed. The thinnest (170 nm) coating consisted mainly of ACP. The coating of approximately 350-nm thick also contained HA, whereas even thicker coatings contained some -TCP in addition to HA. All coatings failed under the scratch test by spalling from the diamond tip; however, the failure load increased as thickness decreased until only plastic deformation and cohesive failure for the thinnest coating was observed (Fernandez-Pradas et al. 2001). Therefore, both the structure and the phase composition of calcium orthophosphate coatings, films and layers might depend on their thickness.

Adhesion

In surgical practice, failure of implants and undesirable tissue responses take place when decohesion of coatings occurs. Therefore, all types of coatings, films and layers must adhere satisfactorily to the underlying substrate irrespective of their intended functions. Generally, the bottom surfaces of the coatings, films and layers are not in the full contact with the substrates. The areas that are in contact are called 'welding points' or 'active zones'. Voids of various shapes and dimensions are located among them. In general, the greater the contact area, the better adhesion of the coating is (Pawlowski 2008). Since the chemical interactions between deposited calcium orthophosphates and substrates are rare, mechanical anchorage is the main mechanism involved in adhesion of calcium orthophosphate coatings, films and layers, in which the substrate surface roughness is the paramount parameter to achieve good adhesion. In many cases of ceramic coatings, the adhesion strength is found to be a linear function of the average surface roughness. Therefore, substrate preparation techniques, such as grit blasting, are used to increase roughness prior to spraying and, hence, increase the adhesion strength (see section 'Brief knowledge on the important pre- and post-deposition procedures'). On the other hand, the amount of mechanical anchorage is reduced if a large amount of shrinkage occurs during solidification of the particles (Heimann 2006). Since the strength of human bones is approximately 18 MPa, all types of coatings, films and layers on the implant surface should have higher or, at least, comparable bond strength. Thus, according to the ISO requirements, the adhesion strength of calcium orthophosphate coatings, films and layers should not be less than 15 MPa (ISO 2000; ISO 2008).

Specifically, the adhesion forces of calcium orthophosphate coatings, films and layers should be high enough to maintain their bioactivity after a surgical implantation. Generally, tensile adhesion testing according to standards ASTM C633 (ASTM C633 2008) and ASTM F-1147-05 (2011) is the most common procedure to determine the quantitative adhesion values to the underlying substrates. Furthermore, fatigue (Surmenev 2012; Mukherjee et al. 2000), scratch (Cheng et al. 2009; Hamdi et al. 2010) and pullout (Cheng et al. 2009) testing, as well as wear resistance (Hamdi et al. 2010), are among the most valuable techniques to provide additional information on the mechanical behavior of calcium orthophosphate coatings, films and layers. Changes in the surface topography can give an indication of wear resistance. For example, coatings with good adherence to the substrate have shown less alteration of its surface roughness, while the study on the different parameters revealed that deposition time was the most influential factor in the wear behavior (Hamdi et al. 2010). The latter was attributed to its correlation with coating thickness. The scratch test is performed with reference to ISO 20502:2005 (ISO 20502 2005). The load at which complete removal of the coating occurs is usually taken as an indication of the adhesion strength. Further details on the mechanical testing methods of calcium orthophosphate coatings, films and layers might be found in literature (Ben-Nissan et al. 2011).

The adhesion strength of calcium orthophosphate coatings, films and layers depends on very many parameters. In the first instance, it strongly depends on the deposition technique. For example, HA coatings, obtained by PLD, showed greater adherence to a titanium alloy when compared with plasma-sprayed HA coatings (Vasanthan et al. 2008). Besides, it might depend on the coating thickness and its chemical composition, namely coatings of 50-μm thick gave higher values of the adhesion strength than those of 240-μm thick (Filiaggi et al. 1991), while scratch tests revealed that the sol-gel-fluorinated HA coating adhered to Ti-alloy substrate up to 35% better as the fluorine concentration increased in the coating (Zhang et al. 2006). Furthermore, the nature, structure and chemical composition of the substrate surface play an important role. For example, a highly roughened substrate surface exhibited higher bond strength as compared to a smooth substrate surface (Nimb et al. 1993). Besides, the adhesion strength of plasma-spayed coatings was found to decrease when either the plate power was reduced (from 28 to 22 kW) or the working distance was increased (from 90 to 130 mm) (Roy et al. 2011). Additionally, the bond strength of calcium orthophosphate coatings deposited on Ti plates pre-treated in an alkali solution followed by heat-treating (600°C for 1 h) in air had a higher value (approximately 35 MPa) if compared to those followed by heat-treating vacuum (approximately 21 MPa). This was attributed to the structural and compositional differences in the interfacial layer of sodium titanates (Wang et al. 2008c). For plasma-assisted deposition techniques of calcium orthophosphates, a good overview on the adhesion strength values of coatings, films and layers is presented in Table 3 of Surmenev (2012).

However, application of various inter-layers (synonym: buffer layers) seems to be the most important way to influence the adhesion strength of calcium orthophosphate coatings, films and layers to diverse substrates. A big number of the available deposition techniques (see 'Preparation' section above), which should be multiplied to a big selection of various substrates, result in a great number of potentially appropriate chemicals to be used as inter-layers between the substrates and calcium orthophosphates. For example, for plasma-assisted deposition methods of calcium orthophosphates, such chemicals as TiO_2 (Rajesh et al. 2011), TiN (Nelea et al. 2000; Yang et al. 2009b, 2009c; Man et al. 2009), ZrO_2 (Nelea et al. 2000) or Al_2O_3 (Nelea et al. 2000), were used as buffer layers. TiO_2 (Nelea et al. 2007; Berezhnaya et al. 2010) and TiN (Nelea et al. 2003) were also used as under-layers for RF magnetron sputtering. Similarly, formation of intermediate layers of titanium hydroxides is required for biomimetic deposition of calcium orthophosphates on Ti (Wang et al. 2008a). To complicate things even further, one should mention, that mutual inter-diffusion of atoms, ions and molecules of calcium orthophosphates from coatings, films and layers from one side and those of a substrate from another side might occur. Especially, this is valid for high temperature deposition techniques; however, the mutual inter-diffusion might happen for any technique at the post-deposition annealing stage (see section 'Brief knowledge on the important pre- and post-deposition procedures'). Various atomic mixed inter-layers are formed as the result. For example, the width (measured by Auger electron spectroscopy) of such inter-layer between a HA coating and magnesium substrate formed by IBAD technique was found to be approximately 3 μm (Yang et al. 2008). Such inter-layers can reduce the mismatch of thermal expansion coefficients between calcium orthophosphates and substrates, or increase the surface area of the material, wettability or heat conductivity; thus, increasing the bonding strength without affecting biocompatibility. Since the subject of inter-layers appears to be very broad, additional details are not specified further.

The adhesion forces depend on various factors, namely for dense coatings under tensile loading, failure usually occurs at the coating/substrate interface because the cohesive strength is higher than the bond strength. For porous coatings, the cohesive strength is low and the fracture occurs inside them (Han et al. 2001). The amorphous coatings have a more brittle nature and less adhesion compared to the crystalline ones (Cleries et al. 2000a). In general, the bond strength of apatite layer to Ti metal substrate is reported to range from 10 to 30 MPa (Kokubo et al. 1996; Kim et al. 1997). Similar values were obtained in another study, where calcium orthophosphate coatings were deposited on Ti substrates by a biomimetic method from two types of SBF. The results indicated that both the ionic concentrations of the SBFs and the surface roughness of the substrates had a significant influence on formation, morphology and bond strength of calcium orthophosphate precipitates. The highest bond strength of the precipitated coatings was about 15.5 MPa (Chen et al.2009b).

Biodegradation

Biodegradation (synonyms: biotic degradation or biotic decomposition) is a chemical dissolution of materials by bacteria and/or other biological means. Since chemical composition of the body fluids might be considered as constancy, biodegradation of calcium orthophosphate coatings, films and layers is controlled by the properties of calcium orthophosphates themselves, which include their chemical composition, Ca/P ratio, crystal structure, crystallinity, porosity, lattice defects, particle sizes and purity (de Groot et al. 1998). For example, the dissolution kinetics of HA layers was studied using the dual constant composition method, and dissolution rates decreased when HA crystallinity increased (Tucker et al. 1996). Similar results were obtained in another study: after implantation, HA coatings with crystallinity of approximately 55% were found to degrade faster and possess better osteoinductivity than those with crystallinity of approximately 98% (Xue et al. 2005). Although biodegradation supposed to be *in vivo* process, various *in vitro* simulations are widely investigated. However, to be closer to the *in vivo* conditions, the biological assessments of calcium orthophosphate coatings, films and layers are performed in various simulating solutions, such as SBF (Kim et al.2010; Chen et al. 2009b; Verestiuc et al. 2004; van der Wal et al. 2005, 2006; Heimann 2009; Łatka et al. 2010; d'Haese et al. 2010; Ntsoane et al. 2011), HBSS (Ueda et al. 2007; Man et al. 2009; Luo et al. 2000), aqueous saline solution (Surmenev et al. 2010; Ueda et al. 2009), Ringer's solution (Gross and Berndt 1994; Gross et al. 1997), phosphate buffered saline (PBS) (Ueda et al. 2007; Boyd et al. 2006; Coelho et al. 2009a), Eagle's minimum essential medium (Lim et al. 2005). Since the simulating solutions often contain dissolved ions of calcium and orthophosphates, both partial dissolution of calcium orthophosphate coatings, films and layers and their re-crystallization occurred (Verestiuc et al. 2004; Heimann 2009; d'Haese et al. 2010; Ntsoane et al. 2011; Lim et al. 2005). For example, as written in the abstract of d,Haese et al. (2010), 'The soaking in SBF homogenizes the morphology of coatings. The sintered zone disappears, and the pores get filled by the reprecipitated calcium phosphates.' One should stress that, due to the presence of other ions in the chemical composition of the aforementioned simulating solutions, in the vast majority of the cases not chemically pure but ion-substituted calcium orthophosphates are precipitated.

Usually, the biodegradation kinetics of calcium orthophosphate coatings, films and layers appears to be proportional to the solubility values of the individual ingredients, listed in Table 1. For example, both bone bonding and bone formation of HA, -TCP and TTCP plasma-sprayed coatings were evaluated by mechanical push-out tests and histological observations after 3, 5, 15 and 28 months of implantation. Among them, -TCP (which was the most soluble phase) showed the most significant degradation after approximately 3 months of implantation, while HA and TTCP showed significant signs of degradation only after approximately 5 months of implantation (Klein et al. 1994a). This resulted in lesser values of the mechanical push-out tests for -TCP-coated implants if compared with those coated by HA and TTCP (Klein et al. 1991). Plasma-spray deposited coatings of HA were found to dissolve faster than the stoichiometric HA did because a high temperature melted HA powder and partly decomposed it into more soluble compounds, such as high-temperature ACP and OA (Pezeshki et al. 2010). Similarly, as-deposited magnetron spattered calcium orthophosphate coatings were almost amorphous (i.e., ACP), and therefore, they completely dissolved after exposure to PBS for only 24 h, while the dissolution rate of the same coatings after annealing (they became crystalline) was found to be more restrained (Boyd et al. 2006). Additionally, HA coatings were found to be less stable than those of FA (Klein et al. 1994b; Dhert et al. 1992; Dhert et al. 1993; Caulier et al. 1995) and of a similar stability with magnesium-whitlockite (i.e., Mg-substituted -TCP) coatings (Dhert et al. 1992, 1993).

On the other hand, there are cases (Gineste et al. 1999; de Bruijn et al. 1994; Cleries et al. 2000b), in which the biodegradation kinetics of calcium orthophosphate coatings, films and layers appeared to be correlated imperfectly with their solubility values (see Table 1). For example, three types of calcium orthophosphate (HA, ACP and -TCP) coatings on titanium alloy substrates, deposited by the laser ablation technique, were immersed in SBF in order to determine their behavior in conditions similar to the human blood plasma. Neither HA nor ACP coatings were found to dissolve in SBF, while a -TCP coating slightly dissolved. Precipitation of an apatitic phase was favored onto both HA and -TCP coatings; however, no precipitation occurred onto ACP coating (Cleries et al. 2000b). Additionally, degradation rates of dental implants with 50- and 100-micron thick coatings of HA, FA and fluorhydroxylapatite (FHA) were studied (Gineste et al. 1999). The implants were inserted in dog jaws and retrieved for histological analysis after 3, 6, and 12 months. The thickness of the calcium orthophosphate coatings was evaluated using an image analysis device. HA and FA coatings (even at 100-micron thickness) were almost totally degraded within the implantation period. In contrast, the FHA coatings did not show significant degradation during the same period (Gineste et al. 1999).

Interaction with Cells and Tissue Responses

The interactions of calcium orthophosphate coatings, films and layers with either cells *in vitro* or surrounding tissues *in vivo* have been studied a lot (Huang et al. 2009; Wang et al. 2004; Choi et al.2003; Ueda et al. 2007; Massaro et al. 2001; Wie et al. 1998; Maistrelli et al. 1993; Hulshoff et al.1996a; Caulier et al. 1997a, 1997b; Antonov et al. 1998; Cleries et al. 2000c; Lo et al. 2000; Jung et al. 2001; Manso et al. 2002c; Heimann et al. 2004; Siebers et al. 2004, 2005; Manders et al. 2006; Simank et al. 2006; Mello et al. 2007; Hashimoto et al. 2008; Coelho and Lemons 2009; Sima et al.2010; Quaranta et al. 2010; Cairns et al. 2010). The *in vitro* trials using different cell lines revealed that in the vast majority of the cases, calcium orthophosphate coatings, films and layers enhanced cellular adhesion, proliferation and differentiation, while the results of the *in vivo* studies revealed that they promoted bone regeneration. For example, a combination of surface geometry and calcium orthophosphate coatings was found to benefit the implant-bone response during the healing phase (Hayakawa et al. 2002). Calcium orthophosphate coatings on titanium implants followed by bisphosphonate-immobilization appeared to be effective in the promotion of osteogenesis on surfaces of dental implants (Yoshinari et al. 2002). A greater percent of bone contact lengths were detected for calcium orthophosphate-coated Ti implants compared with control Ti implants 3 and 12 weeks after implant placement (Ong et al. 2002). Similar results were obtained in other studies (Dalton and Cook 1995; Nguyen et al. 2004; Yan et al. 2006).

Concerning the experiments with cells, human gingival fibroblasts attachment, spreading, extracellular matrix production and focal adhesion plaque formation were investigated on commercially pure Ti, HA-coated Ti and porous TCP/HA-coated Ti. TCP/HA and HA coatings exhibited that both the attached cell number and cell spreading area were higher than that on pure Ti and focal adhesion plaque formed earlier than that of uncoated substrate. The attached cell number and type I collagen formation on TCP/HA coatings were more than that on HA ones (Zhao et al. 2005). Osteoblasts were successfully grown on the surface of OCP (Bigi et al. 2005b) and HA (Cao et al.2010a; Bigi et al. 2005b; Ball et al. 2001); both types of coatings, layers and films were found to favor osteoblast proliferation, activation of their metabolism and differentiation. Furthermore, the *in vitro* cell-culture studies using MG63 osteoblast-like cells were performed on calcium orthophosphate coatings deposited on titanium by plasma spray, sol-gel and sputtering techniques. The study demonstrated the ability of cells to proliferate on the materials tested. The sol-gel coating was found to promote higher cell growth, greater alkaline phosphatase activity and greater osteocalcin production compared to the sputtered and plasma-sprayed coatings (Massaro et al. 2001). In another study, calcium orthophosphate coatings were found to induce significantly higher cell differentiation levels than the uncoated control (Bucci-Sabattini et al. 2010).

A study by Cairns et al. (2010) should be described especially. Calcium orthophosphate thin films were deposited onto substrates with varying topography. Then, a layer of fibronectin was deposited from solution onto each surface, and the response of MG63 osteoblast-like cells was studied. The results revealed that, in all cases, the presence of the adsorbed fibronectin layer improved cell adhesion, proliferation and promoted early onset differentiation. Moreover, the nature and scale of the response appeared to be influenced by the surface topography of the substrates. Specifically, cells on the fibronectin-coated calcium orthophosphate thin films with regular topographical features in the nanometer range showed statistically significant differences in focal adhesion assembly, osteocalcin expression and alkaline phosphase activity compared to the calcium orthophosphate films without those topographical features (Cairns et al. 2010). Therefore, both an adsorbed bioorganics and a surface topography of the substrates appear to influence cell adhesion and differentiation.

In vivo results correlate well with the *in vitro* ones. Osseointegration rates of porous-surfaced Ti6Al4V implants with control (unmodified sintered coatings) were compared to porous-surfaced implants modified through the addition of either an inorganic or organic route sol-gel-formed calcium orthophosphate films. Implants were placed in distal femoral rabbit condyle sites and, following a 9-day healing period, implant fixation strength was evaluated using a pullout test. Both types of calcium orthophosphate films significantly enhanced the early rate of bone in-growth and fixation as evidenced by higher pullout force and interface stiffness compared with controls. However, there was no significant difference between calcium orthophosphate-coated implants prepared using the two different methods (Gan et al. 2004).

To conclude this section, one should note that the positive clinical benefits of calcium orthophosphate coatings, layers and films were not always detected. For example, a study was undertaken to evaluate the processes involved in biological responses of the Ti-6Al-7Nb alloy with and without HA coatings with both *in vitro* and *in vivo* tests. The results with HA coating appeared to be similar to those obtained on the uncoated samples (Lavos-Valereto et al. 2001). Similarly, neither positive nor negative influence of the presence of HA coatings on the surface of implants was detected during the 10-year (Lazarinis et al. 2011), 13-year (Camazzola et al. 2009), 15-year (Stilling et al. 2009) and undisclosed (Lee et al. 2000; Gandhi et al. 2009) follow-ups. Besides, there are cases in which short-term (4 weeks) advantages of the calcium orthophosphate-coated implants were found in animal studies, whereas no significant differences to the uncoated samples were found after 6 months (Gottlander et al. 1997a). Furthermore, inflammatory tissue reaction cases have been detected (Piattelli et al. 1995; Walschus et al. 2009). Interestingly, the short-term inflammatory response against a HA coating on Ti was lower in comparison to a DCPD coating on Ti. The observed differences between the Ti-DCPD implants and the Ti-HA implants were attributed to their dissolution characteristics:

the HA coating on Ti showed increased stability and, hence reduced the inflammatory response (Walschus et al. 2009). Furthermore, HA coatings were found to be a risk factor for cup revision due to aseptic loosening (Lazarinis et al. 2010). Thus, precautions to prevent contamination (asepsis) and/or infection (perioperative antibiotics) appear to be more important for the calcium orthophosphate-coated implants if compared with the uncoated ones (Oosterbos et al. 2002).

Biomedical Applications

Already in 1987, de Groot et al. (1987), published a paper on the development of plasma-sprayed HA coatings on metallic implants. The same year the same researchers published the results of the first clinical study (Geesink et al. 1987). Shortly afterwards, Furlong and Osborn, two leading surgeons in the orthopedics field, began implanting plasma-sprayed HA stems in patients (Furlong and Osborn 1991) followed by other clinicians (Bauer et al. 1991; Buma and Gardeniers 1995). Since then, plentiful reports have been published about the biomedical advantages of such coated implants. To summarize the available information on the biomedical and biomechanical properties of implants coated by calcium orthophosphates, one can claim the following: If compared to uncoated implants, the presence of calcium orthophosphate deposits were found to induce bone contacts to the implants (Dhert et al. 1992, 1993; Thomas et al. 1989; Jansen et al. 1991; Gottlander et al. 1997b; Hulshoff and Jansen 1997; Hayakawa et al. 2000; Mohammadi et al. 2004; Park et al. 2005; Siebers et al. 2007; Kuroda et al. 2007; Chae et al. 2008; Schwarz et al. 2009; Junker et al. 2010; Suzuki et al. 2010); improve implant fixation (Yang et al. 1997; Søballe et al. 1993; Daugaard et al. 2010); show higher torque values (Park et al. 2005; Junker et al. 2010; Granato et al. 2009) and push-out strength (Ozeki et al. 2001); facilitate bridging of small gaps between implants and surrounding bones (Søballe et al. 1991; Stephenson et al. 1991), reduce metal ion release from the metallic substrates (Surmenev et al. 2010; Ducheyne and Healy 1988; Sousa and Barbosa 1996; Ozeki et al. 2003); slow down metal degradation and/or its corrosion (Metikoš-Hukovi et al. 2003; Yang et al. 2008; Cheng and Roscoe 2005); accelerate bone growth (Cook et al. 1992; Wang et al. 2009), remodeling (Pilliar et al. 1991; Yoon et al. 2009) and osteointegration rate (Bigi et al. 2008; Lee et al. 2011); induce osteoconductivity (Cao et al. 2010b), improve the early bone (Yang et al. 1996; Mohammadi et al. 2003) and healing (Vercaigne et al. 2000b) responses; and result in lack of formation of fibrous tissues (Figure 11) (Layrolle 2011; Dostálová et al. 2001), as well as increase the clinical performance of orthopedic hip systems (see below). In addition, calcium orthophosphate coatings, films and layers might be used for incorporation of drugs and important biologically active compounds, such as peptides, hormones and growth factors (Siebers et al. 2006). In the case of porous implants, calcium orthophosphate coatings enhance bone ingrowth into the pores (Suchanek and Yoshimura 1998). Furthermore, studies concluded that there was significantly less pin loosening in calcium orthophosphate-coated groups (Saithna 2010). Thus, the majority of the clinical studies are optimistic about the *in vivo* performance of calcium orthophosphate-coated prostheses. However, to be objective, one must mention on the studies in which no positive biomedical and/or biomechanical effects of calcium orthophosphate coatings, films and layers have been detected (Tieanboon et al. 2009; Coelho et al. 2009b). Besides, the presence or absence of the positive biomedical and/or biomechanical effects of calcium orthophosphate coatings, films and layers might depend on the deposition technique used (Hulshoff et al. 1996b, 1997), as well as on the coating vendor (Dalton and Cook 1995). These uncertainties might be due to several reasons, such as variability in chemical and phase composition, porosity, admixtures, *etc.*

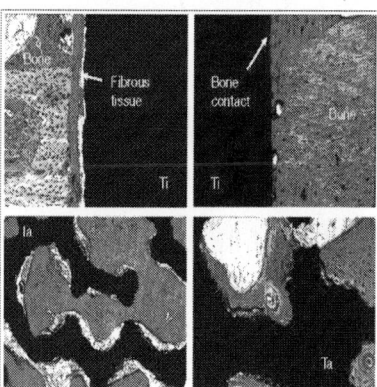

Figure 11: Comparison of bone-integrative properties. Non-coated (left) and biomimetically coated by calcium orthophosphates metal implants (right) af-

ter implantation in the femur of goats for 6 weeks. Reprinted fromLayrolle (2011) with permission.

In biomedical applications, bone grafts are usually much thicker than coatings, films or layers applied to them. Nevertheless, the coated implants combine the surface biocompatibility and bioactivity of calcium orthophosphates with the core strength of strong substrates (Figure 12). The clinical results for calcium orthophosphate-coated implants reveal that they have much longer life times after implantation than uncoated devices, and therefore, they are particularly beneficial for younger patients (Capello et al. 1997). Their biomedical properties are approaching those of bioactive glass-coated implants (Wheeler et al. 2001; Mistry et al. 2011).

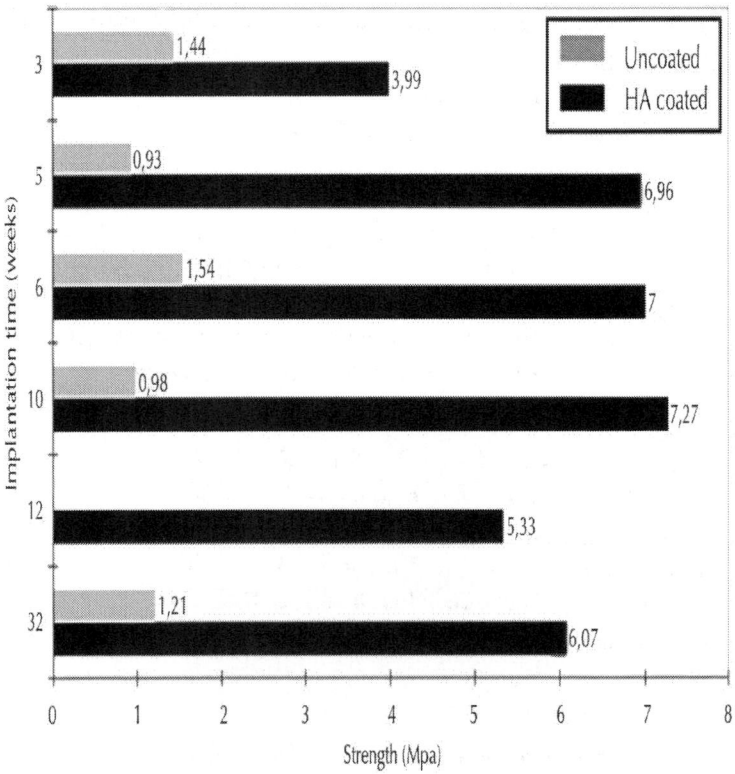

Figure 12: Time-dependent plasma-sprayed HA coating. This figure shows how a plasma-sprayed HA coating on a porous titanium (dark bars) dependent on the implantation time will improve the interfacial bond strength com-

pared to uncoated porous titanium (light bars). Reprinted from Hench (1991) with permission.

Since, among calcium orthophosphates, HA is the most popular material to be deposited as coatings, films and layers, the vast majority of the clinical investigations was performed with HA. For example, HA coating as a system of fixation of hip implants *in vivo* was found to work well in the short to medium terms (2 years (Geesink 1990), 6 years (Geesink and Hoefnagels 1995), 8 years (Wheeler 1996; Chang et al. 2006), 9 to 12 years (MaNally et al. 2000), 10 years (Oosterbos et al. 2004; Trisi et al. 2005), 10 to 15.5 years (Matsumine et al. 2004), 10 to 17 years (Muirhead-Allwood et al. 2010), 13 to 15 years (Shetty et al. 2005), 15 to 21 years (Rajaratnam et al. 2008), 16 years (Buchanan 2005), 17 years (Buchanan 2006) and 19 years (Buchanan and Goodfellow 2008)). In 2004, a special book summarizing the studies with HA-coated implants and the 'state of the art' of HA coatings in orthopedics at the close of 2002 was published (Epinette and Manley 2004). Similar data for HA-coated dental implants are also available (Tinsley et al. 2001; Binahmed et al. 2007; Iezzi et al.2007). Nevertheless, even longer-term clinical results are awaited with a great interest. The biomedical aspects of osteoconductive coatings for total joint arthroplasty have been reviewed elsewhere (Geesink 2002). Additional details on calcium orthophosphate coatings, films and layers might be found in excellent reviews (Narayanan et al. 2010; Paital and Dahotre 2009a; León andJansen 2009).

Nevertheless, one must stress that although many experiments concerning the *in vivo* studies of calcium orthophosphate coatings, films and layers have indicated a stronger and faster fixation, as well as more bone ingrowth at the interface, the clinical performance of such coatings, films and layers is still far from the perfection. Some of the major concerns associated with the usage of calcium orthophosphate coatings, films and layers in actual body environment, with regard to their long-term stability can be listed as follows (Hulshoff et al. 1996a; Caulier et al. 1997a, 1997b):

- The degradation and resorption of calcium orthophosphate coatings, films and layers in a biological environment could lead to disintegration of the coating, resulting in the loss of both coating-substrate bond strength and the implant fixation.

- Coating delamination and disintegration with the formation of particulate debris are also major concern
- Calcium orthophosphate coatings, films and layers may also lead to increased polyethylene wear from the acetabular cup and, thereby, alleviate the problem of osteolysis.

In spite of the long history and the aforementioned achievements, still not all concerns on the surgical applications of calcium orthophosphate coatings, films and layers have been eliminated. Still, a limited amount of the *in vivo* studies is available in the literature. The limitations to such experiments may be attributed to any of the following reasons:

- Difficulty in selection of a suitable animal model to simulate the actual mechanical loading and unloading conditions the implant might undergo in a human body environment.
- The need to sacrifice a large number of animals, since most of these experiments demand a statistical analysis to validate the results.
- A high cost and a long period of clinical testing these experiments demand.
- Lack of coordination among material scientists and biologists and thereby an insufficient understanding of this interdisciplinary subject.
- Serious ethical concerns on the use of animals for experimental studies as they are subjected to painful procedures or toxic exposures during the course of test.

To conclude this section, one must note the following: Even though the importance and the need for development of calcium orthophosphate coatings, films and layers have been recognized, it is still mostly being explored on research level, and after extensive search of open literature, these coatings appear to have made limited headway into commercialization. In spite of mention of the commercial products such as hip and dental implants produced by Zimmer Orthopedics (Freiburg, Germany), Smith and Nephew (Memphis, TN, USA), and Biomet, the science and technology related to their manufacturing is not disclosed by any one of them due to the proprietary reasons. Hence, at this point it is difficult to bring a detailed discussion on commercialization of calcium orthophosphate coatings, films and layers (Paital and Dahotre 2009a).

Future Directions

A potential drawback of the majority of the deposition techniques of calcium orthophosphate coatings, films and layers is their relatively high cost for a large scale production. Therefore, to decrease processing time and make their manufacturing commercially viable, it is desirable to process the thinnest coating that would significantly increase the biological response (Coelho et al. 2009c). Much attention should be paid to functionally graded structures with an amorphous top layer and a crystalline layer underneath (Wang and Zreiqat 2010). This allows adjusting the coating resorption rates to the values at which new bone grows at early stages when it is of the most importance for the bone mineralization process. Furthermore, therapeutic capabilities of calcium orthophosphate coatings, films and layers as templates for the *in situ* delivery of drugs and osteoinductive agents (peptides, hormones and growth factors) at the required times should be elucidated much better.

CONCLUSIONS

Solid implants prepared from various materials often possess a poor biocompatibility with a simultaneous lack of the osteogenic properties in order to promote bone healing. In addition, direct bone-to-implant contacts are desired for a biomechanical anchoring of implants rather than fibrous tissue encapsulation. All these problems might be solved by applying calcium orthophosphate coatings, films and layers. The aim is to provide the implants with surface biological properties for adsorption of proteins, adhesion of cells and bone apposition. Therefore, the available knowledge on calcium orthophosphate and, most notably, HA coatings, films and layers on various substrates has been summarized in this review. Since all available deposition techniques have both advantages and shortcomings of their own (Table 3), still there are no standard guidelines for depositing calcium orthophosphates on the implant surfaces. In general, dissolution of calcium orthophosphate coatings, films and layers improves implant osseointegration and is the basic requirement for bioactivity. However, this dissolution diminishes the stability and increases the potential for loosening of the implants. Calcium orthophosphate coatings, films and layers of lower solubility and higher stability are desirable for the long-term performance of implants because they promote faster initial bone fixation, bridge larger gaps in the misfit and degrade at a controlled rate.

Although animal and *in vitro* studies have already reported on the benefits of using calcium orthophosphate-coated implants, as well as the risks of dissolution, the short-term studies did not demonstrate that the dissolution of calcium orthophosphate coatings, films and layers led to a loss of implants. In addition, many *in vivo* and clinical studies did not consider the chemical and structural characterizations of the coatings. Under these conditions, any comparisons among various reports and studies are difficult.

New promising techniques for coating medical devices are continuously investigated. Future investigations on various coating processes will have to include clinical trials to get better understanding of bone responses to coated-implant surfaces, as well as studies on coupling of calcium orthophosphate coatings, layers and films with drugs, growth factors and cells. Although it has been generally accepted that calcium orthophosphate coatings, layers and films improve bone strength and initial osteointegration rate, the optimal coating properties required to achieve maximal bone response are yet to be reported. As such, the use of well-characterized calcium orthophosphate coatings, layers and films cell culture studies, animal studies and clinical studies should be well documented to avoid controversial results.

In addition, the clinicians need to take into consideration the enhanced bacterial susceptibility of calcium orthophosphate-coated implants if compared to the metallic ones. Besides, the clinicians need to consider possible failures of calcium orthophosphate coatings, films and layers as a result of coating-substrate interfacial fracture. It is also important that the clinical investigators be well versed with the material characterizations of the coated implants.

REFERENCES

1.   Ali MY, Hung W, Yongqi F (2010) A review of focused ion beam sputtering. Int. J. Precision Eng. Manuf 11:157-170

2.   Amjad Z (1997) Calcium phosphates in biological and industrial systems. Kluwer, Boston, MA, USA. p 529

3.   Amy RL, Storb R (1955) Selective mitochondrial damage by a ruby laser microbeam: an electron microscopic study. Science 122:756-758

4.   Antonov EN, Bagratashvili VN, Popov VK, Sobol EN, Howdle SM, Joiner C, Parker KG, Parker TL, Doctorov AL, Likhanov VB, Volozhin AI, Alimpiev SS, Nikiforov SM (1998) Biocompatibility of laser-deposited hydroxyapatite coatings on titanium and polymer implant materials. J Biomed Opt 3:423-428

5.   Arias JL, García-Sanz FJ, Mayor MB, Chiussi S, Pou J, León B, Pérez-Amor M (1998) physicochemical properties of calcium phosphate coatings produced by pulsed laser deposition at different water vapour pressures. Biomaterials 19:883-888

6.   ASTM International (2008) ASTM C633 - 01 Standard test method for adhesion or cohesion strength of thermal spray coatings.http://www.astm.org/Standards/C633.htm

7.   ASTM International (2011) ASTM F1147 - 05 Standard test method for tension testing of calcium phosphate and metallic coatings.http://www.astm.org/Standards/F1147.htm

8.   Ball MD, Downes S, Scotchford CA, Antonov EN, Bagratashvili VN, Popov VK, Lo WJ, Grant DM, Howdle SM (2001) Osteoblast growth on titanium foils coated with hydroxyapatite by pulsed laser ablation. Biomaterials 22:337-347

9.   Ban S, Maruno S (1998) Hydrothermal-electrochemical deposition of hydroxyapatite. J Biomed Mater Res 42:387-395

10.   Bao Q, Chen C, Wang D, Lei T, Liu J (2006) Pulsed laser deposition of hydroxyapatite thin films under Ar atmosphere. Mater Sci Eng, A 429:25-29

11.   Barrere F, van Blitterswijk CA, de Groot K, Layrolle P (2002) Influence of ionic strength and carbonate on the Ca-P coating formation from SBF × 5 solution. Biomaterials 23:1921-1930

12.   Barrere F, van Blitterswijk CA, de Groot K, Layrolle P (2002) Nucleation of biomimetic Ca-P coatings on Ti6Al4V from a SBF × 5 solution: influence of magnesium. Biomaterials 23:2211-2220

13.   Barrere F, Snel MME, van Blitterswijk CA, de Groot K, Layrolle P (2004) Nano-scale study of the nucleation and growth of calcium phosphate coating on titanium implants. Biomaterials 25:2901-2910

14. Barthell BL, Archuleta TA, Kossowsky R (1989) Ion beam deposition of calcium hydroxyapatite. Mater Res Soc Symp Proc 110:709-715

15. Bauer TW, Geesink RGT, Zimmerman R, McMahon JT (1991) Hydroxyapatite-coated femoral stems. Histological analysis of components retrieved at autopsy. J. Bone Joint Surg. A 73:1439-1452

16. Ben-Nissan B, Latella BA, Bendavid A (2011) 3.305. Biomedical thin films: mechanical properties. In: Ducheyne P, Healy K, Hutmacher DW, Grainger DW, Kirkpatrick CJ (eds) Comprehensive biomaterials, Elsevier, Amsterdam, Netherlands. pp 63-73 Vol. 3.

17. Berezhnaya AY, Mittova VO, Kukueva EV, Mittova IY (2010) Effect of high-temperature annealing on solid-state reactions in hydroxyapatite/TiO2 films on titanium substrates. Inorg Mater 46:971-977

18. Besra L, Liu M (2007) A review on fundamentals and applications of electrophoretic deposition (EPD). Prog Mater Sci 52:1-61

19. Bigi A, Boanini E, Bracci B, Facchini A, Panzavolta S, Segatti F, Struba L (2005) Nanocrystalline hydroxyapatite coatings on titanium: a new fast biomimetic method. Biomaterials 26:4085-4089

20. Bigi A, Bracci B, Cuisinier F, Elkaim R, Fini M, Mayer I, Mihailescu IN, Socol G, Sturba L, Torricelli P (2005) Human osteoblast response to pulsed laser deposited calcium phosphate coatings. Biomaterials 26:2381-2389

21. Bigi A, Fini M, Bracci B, Boanini E, Torricelli P, Giavaresi G, Aldini NN, Facchini A, Sbaiz F, Giardino R (2008) The response of bone to nanocrystalline hydroxyapatite-coated Ti13Nb11Zr alloy in an animal model. Biomaterials 29:1730-1736

22. Bigi A, Boanini E, Capuccini C, Fini M, Mihailescu IN, Ristoscu C, Sima F, Torricelli P (2009) Biofunctional alendronate-hydroxyapatite thin films deposited by matrix assisted pulsed laser evaporation. Biomaterials 30:6168-6177

23. Binahmed A, Stoykewych A, Hussain A, Love B, Pruthi V (2007) Long-term follow-up of hydroxyapatite-coated dental implants - a clinical trial. Int J Oral Max Impl 22:963-968

24. Bini RA, Santos ML, Filho EA, Marques RFC, Guastaldi AC (2009) Apatite coatings onto titanium surfaces submitted to laser ablation with different energy densities. Surf CoatTechnol 204:399-403

25. Blackwood DJ, Seah KHW (2009) Electrochemical cathodic deposition of hydroxyapatite: improvements in adhesion and crystallinity. Mater Sci Eng C 29:1233-1238

26. Blalock T, Bai X, Rabiei A (2007) A study on microstructure and properties of calcium phosphate coatings processed using ion beam assisted deposition on heated substrates. Surf CoatTechnol 201:5850-5858

27. Boanini E, Torricelli P, Fini M, Sima F, Serban N, Mihailescu IN, Bigi A (2012) Magnesium and strontium doped octacalcium phosphate thin films by matrix assisted pulsed laser evaporation. J Inorg Biochem 107:65-72

28. Bocanegra-Bernal MH (2004) Hot isostatic pressing (HIP) technology and its applications to metals and ceramics. J Mater Sci 39:6399-6420

29. Boyd AR, Meenan BJ, Leyland NS (2006) Surface characterisation of the evolving nature of radio frequency (RF) magnetron sputter deposited calcium phosphate thin films after exposure to physiological solution. Surf CoatTechnol 200:6002-6013

30. Boyd AR, Duffy H, McCann R, Cairns ML, Meenan BJ (2007) The influence of argon gas pressure on co-sputtered calcium phosphate thin films. Nucl Instrum Methods Phys Res B 258:421-428

31. Brès E, Hardouin P (eds) (1998) Les matériaux en phosphate de calcium: aspects fondamentaux Sauramps Medical, Montpellier, France. p 176

32. Brinker CJ, Frye GC, Hurd AJ, Ashley CS (1991) Fundamentals of sol–gel dip coating. Thin Solid Films 201:97-108

33. Brown PW, Constantz B (eds) (1994) Hydroxyapatite and related materials CRC Press, Boca Raton, FL, USA. p 343

34. Bucci-Sabattini V, Cassinelli C, Coelho PG, Minnici A, Trani A, Ehrenfest DMD (2010) Effect of titanium implant surface nanoroughness and calcium phosphate low impregnation on bone cell activity in vitro. Oral Surg Oral Med Oral Pathol Oral Radiol Endodontol 109:217-224

35. Buchanan JM (2005) 16 year review of hydroxyapatite ceramic coated hip implants – a clinical and histological evaluation. Key Eng. Mater. 284–286:1049-1052

36. Buchanan JM (2006) 17 year review of hydroxyapatite ceramic coated hip implants – a clinical and histological evaluation. Key Eng. Mater. 309–311:1341-1344

37. Buchanan JM, Goodfellow S (2008) Nineteen years review of hydroxyapatite ceramic coated hip implants: a clinical and histological evaluation. Key Eng Mater 361–363:1315-1318

38. Buma P, Gardeniers JW (1995) Tissue reactions around a hydroxyapatite-coated hip prostheses: case report of a retrieveal specimen. J Arthroplasty 10:389-395

39. Burgess AV, Story BJ, La D, Wagner WR, LeGeros JP (1999) Highly crystalline MP-1 hydroxylapatite coating. Part I: In vitro characterization and comparison to other plasma-sprayed hydroxylapatite coatings. Clin Oral Implant Res 10:245-256

40. Cairns ML, Meenan BJ, Burke GA, Boyd AR (2010) Influence of surface topography on osteoblast response to fibronectin coated calcium phosphate thin films. Coll. Surf. B 78:283-290

41. Callahan TJ, Gantenberg JB, Sands BE (1994) Calcium phosphate (Ca-P) coating draft guidance for preparation of Food and Drug Administration (FDA) submissions for orthopedic and dental endosseous implants. In: Horowitz E, Parr JE (eds) Characterization and performance of calcium phosphate coatings for implants, ASTM STP 1196, Philadelphia, PA, USA. pp 185-197

42. Camazzola D, Hammond T, Gandhi R, Davey JR (2009) A randomized trial of hydroxyapatite-coated femoral stems in total hip arthroplasty. A 13-year follow-up. J Arthroplasty 24:33-37

43. Campbell AA (2003) Bioceramics for implant coatings. Mater Today 6:26-30

44. Cannillo V, Lusvarghi L, Sola A, Barletta M (2009) Post-deposition laser treatment of plasma sprayed titania-hydroxyapatite functionally graded coatings. J Eur Ceram Soc 29:3147-3158

45. Cao Y, Weng J, Chen J, Feng J, Yang Z, Zhang X (1996) Water vapor-treated hydroxyapatite coatings after plasma spraying and their characteristics. Biomaterials 17:419-424

46. Cao N, Dong J, Wang Q, Ma Q, Wang F, Chen H, Xue C, Li M (2010) Plasma-sprayed hydroxyapatite coating on carbon/carbon composite scaffolds for bone tissue engineering and related tests in vivo. J Biomed Mater Res Am 92A:1019-1027

47. Cao N, Dong J, Wang Q, Ma Q, Xue C, Li M (2010) An experimental bone defect healing with hydroxyapatite coating plasma sprayed on carbon/carbon composite implants. Surf CoatTechnol 205:1150-1156

48. Capello WD, D'Antonio JA, Feinberg JR, Manley MT (1997) Hydroxyapatite-coated total hip femoral components in patients less than fifty years old. Clinical and radiographic results after five to eight years of follow-up. J Bone Joint Surg Am 79:1023-1029

49. Carradó A (2010) Structural, microstructural, and residual stress investigations of plasma-sprayed hydroxyapatite on Ti-6Al-4 V. ACS Appl Mater Interf 2:561-565

50. Carradò A, Viart N (2010) Nanocrystalline spin coated sol–gel hydroxyapatite thin films on Ti substrate: towards potential applications for implants. Solid State Sci 12:1047-1050

51. Caulier H, van der Waerden JPCM, Paquay YCGJ, Wolke JGC, Kalk W, Naert I, Jansen JA (1995) Effect of calcium phosphate (Ca-P) coatings on trabecular bone response: a histological study. J Biomed Mater Res 29:1061-1069

52. Caulier H, van der Waerden JPCM, Wolke JGC, Kalk W, Naert I, Jansen JA (1997) A histological and histomorphometrical evaluation of the application of screw-designed calciumphosphate (Ca-P)-coated implants in the cancellous maxillary bone of the goat. J Biomed Mater Res 35:19-30

53. Caulier H, Hayakawa T, Naert I, van der Waerden JPCM, Wolke JGC, Jansen JA (1997) An animal study on the bone behaviour of Ca-P-coated implants: influence of implant location. J Mater Sci Mater Med 8:531-536

54. Chae GJ, Jung UW, Jung SM, Lee IS, Cho KS, Kim CK, Choi SH (2008) Healing of surgically created circumferential gap around nano-coating surface dental implants in dogs. Surf. Interf. Analysis 40:184-187

55. Chang JK, Chen CH, Huang KY, Wang GJ (2006) Eight-year results of hydroxyapatite-coated hip arthroplasty. J Arthroplasty 21:541-546

56. Cheang P, Khor KA (1995) Thermal spraying of hydroxyapatite (HA) coatings: effects of powder feedstock. J Mater Process Technol 48:429-436

57. Chen XB, Li YC, Plessis JD, Hodgson PD, Wen C (2009) Influence of calcium ion deposition on apatite-inducing ability of porous titanium for biomedical applications. Acta Biomater 5:1808-1820

58. Chen X, Li Y, Hodgson PD, Wen C (2009) Microstructures and bond strengths of the calcium phosphate coatings formed on titanium from different simulated body fluids. Mater Sci Eng C 29:165-171

59. Cheng X, Roscoe SG (2005) Corrosion behavior of titanium in the presence of calcium phosphate and serum proteins. Biomaterials 26:7350-7356

60. Cheng GJ, Ye C (2010) Experiment, thermal simulation, and characterizations on transmission laser coating of hydroxyapatite on metal implant. J Biomed Mater Res A 92A:70-79

61. Cheng K, Ren C, Weng W, Du P, Shen G, Han G, Zhang S (2009) Bonding strength of fluoridated hydroxyapatite coatings: a comparative study on pull-out and scratch analysis. Thin Solid Films 517:5361-5364

62. Choi JM, Kim HE, Lee IS (2000) Ion-beam-assisted deposition (IBAD) of hydroxyapatite coating layer on Ti-based metal substrate. Biomaterials 21:469-473

63. Choi J, Bogdanski D, Koller M, Esenwein SA, Muller D, Muhr G, Epple M (2003) Calcium phosphate coating of nickel-titanium shape-memory alloys. Coating procedure and adherence of leukocytes and platelets. Biomaterials 24:3689-3696

64. Chou BY, Chang E (2001) Interface investigation of plasma-sprayed hydroxyapatite coating on titanium alloy with ZrO2 intermediate layer as bond coat. Scr Mater 45:487-493

65. Chow LC, Eanes ED (eds) Octacalcium phosphate (Monographs in oral science). vol. 18 S Karger Pub, Basel, Switzerland. p 168

66. Cizek J, Khor KA, Prochazka Z (2007) Influence of spraying conditions on thermal and velocity properties of plasma sprayed hydroxyapatite. Mater Sci Eng C 27:340-344

67. Cleries L, Martinez E, Fernandez-Pradas JM, Sardin G, Esteve J, Morenza JL (2000) Mechanical properties of calcium phosphate coatings deposited by laser ablation. Biomaterials 21:967-971

68. Cleries L, Fernandez-Pradas JM, Morenza JL (2000) Behaviour in simulated body fluid of calcium phosphate coatings obtained by laser ablation. Biomaterials 21:1861-1865

69. Cleries L, Fernandez-Pradas JM, Morenza JL (2000) Bone growth on and resorption of calcium phosphate coatings obtained by pulsed laser deposition. J Biomed Mater Res 49:43-52

70. Coelho PG, Lemons JE (2009) Physico/chemical characterization and in vivo evaluation of nanothickness bioceramic depositions on alumina-blasted/acid-etched Ti-6Al-4 V implant surfaces. J Biomed Mater Res A 90A:351-361

71. Coelho PG, de Assis SL, Costa I, Thompson VP (2009) Corrosion resistance evaluation of a Ca- and P-based bioceramic thin coating in Ti-6Al-4 V. J Mater Sci Mater Med 20:215-222

72. Coelho PG, Cardaropoli G, Suzuki M, Lemons JE (2009) Early healing of nanothickness bioceramic coatings on dental implants. An experimental study in dogs. J Biomed Mater Res B Appl Biomater 88B:387-393

73. Coelho PG, Granjeiro JM, Romanos GE, Suzuki M, Silva NRF, Cardaropoli G, van Thompson P, Lemons JE, Coelho PG, Granjeiro JM, Romanos GE, Suzuki M, Silva NRF, Cardaropoli G, van Thompson P, Lemons JE (2009) Basic research methods and current trends of dental implant surfaces. J Biomed Mater Res B Appl Biomater 88B:579-596

74. Cook SD, Thomas KA, Kay JF, Jarcho M (1988) Hydroxyapatite-coated titanium for orthopedic implant applications. Clin. Orthop. Rel. Res. 232:225-243

75. Cook SD, Thomas KA, Dalton JE, Volkman TK, Whitecloud TS III, Kay JF (1992) Hydroxylapatite coating of porous implants improves bone ingrowth and interface attachment strength. J Biomed Mater Res 26:989-1001

76. Cooley DR, van Dellen AF, Burgess JO, Windeler S (1992) The advantages of coated titanium implants prepared by

radiofrequency sputtering from hydroxyapatite. Prosthetic Dent. 67:93-100

77. Cotell CM (1993) Pulsed laser deposition and processing of biocompatible hydroxylapatite thin films. Appl Surf Sci 69:140-148

78. Cotell CM, Chrisey DB, Grabowski KS, Sprague JA, Gosset CR (1992) Pulsed laser deposition of hydroxylapatite thin films on Ti–6Al–4 V. J Appl Biomed 3:87-93

79. Craciun V, Boyd IW, Craciun D, Andreazza P, Perriere J (1999) Vacuum ultraviolet annealing of hydroxyapatite films grown by pulsed laser deposition. J Appl Phys 85:8410-8414

80. Cuerno R, Barabási AL (1995) Dynamic scaling of ion-sputtered surfaces. Phys Rev Lett 74:4746-4749

81. Cui FZ, Luo ZS, Feng QL (1997) Highly adhesive hydroxyapatite coatings on titanium alloy formed by ion beam assisted deposition. J Mater Sci Mater Med 8:403-405

82. D'Haese R, Pawlowski L, Bigan M, Jaworski R, Martel M (2010) Phase evolution of hydroxapatite coatings suspension plasma sprayed using variable parameters in simulated body fluid. Surf CoatTechnol 204:1236-1246

83. Dalton JE, Cook SD (1995) In vivo mechanical and histological characteristics of HA-coated implants vary with coating vendor. J Biomed Mater Res 29:239-245

84. Daugaard H, Elmengaard B, Bechtold JE, Jensen T, Soballe K (2010) The effect on bone growth enhancement of implant coatings with hydroxyapatite and collagen deposited electrochemically and by plasma spray. J Biomed Mater Res A 92A:913-921

85. De Andrade MC, Filgueiras MRT, Ogasawara T (2002) Hydrothermal nucleation of hydroxyapatite on titanium surface. J Eur Ceram Soc 22:505-510

86. De Bruijn JD, Bovell YP, van Blitterswijk CA (1994) Structural arrangements at the interface between plasma sprayed calcium phosphates and bone. Biomaterials 15:543-550

87. De Groot K, Geesink RGT, Klein CPAT, Serekian P (1987) Plasma sprayed coatings of hydroxylapatite. J Biomed Mater Res 21:1375-1381

88. De Groot K, Wolke JGC, Jansen JA (1998) Calcium phosphate coatings for medical implants. Proc Inst Mech Eng Part J Eng Med 212:137-147

89. De Sena LA, de Andrade MC, Rossi AM, Soares GDA (2002) Hydroxypatite deposiiton by electrophoresis on titanium sheets with different surface finishing. J Biomed Mater Res 60:1-7

90. Dey A, Mukhopadhyay AK (2010) Anisotropy in nanohardness of microplasma sprayed hydroxyapatite coating. Adv Appl Ceram 109:346-354

91. Dey A, Mukhopadhyay AK (2011) Fracture toughness of microplasma-sprayed hydroxyapatite coating by nanoindentation. Int J Appl Ceram Technol 8:572-590

92. Dey A, Nandi SK, Kundu B, Kumar C, Mukherjee P, Roy S, Mukhopadhyay AK, Sinha MK, Basu D (2011) Evaluation of hydroxyapatite and β-tri calcium phosphate microplasma spray coated pin intra-medullary for bone repair in a rabbit model. Ceram Int 37:1377-1391

93. Dhert WJA, Klein CPAT, Wolke JGC, van der Velde EA, de Groot K, Rozing PM (1992) A mechanical investigation of fluorapatite, magnesium whitlockite and hydroxylapatite plasma-sprayed coatings in goats. J Biomed Mater Res 25:1183-1200

94. Dhert WJA, Klein CPAT, Jansen JA, van der Velde EA, Vriesde RC, de Groot K, Rozing PM (1993) A histological and histomorphometrical investigation of fluorapatite, magnesium whitlockite and hydroxylapatite plasma-sprayed coatings in goats. J Biomed Mater Res 27:127-138

95. Dinda GP, Shin J, Mazumder J (2009) Pulsed laser deposition of hydroxyapatite thin films on Ti-6Al-4 V: effect of heat treatment on structure and properties. Acta Biomater 5:1821-1830

96. Ding SJ (2003) Properties and immersion behavior of magnetron-sputtered multi-layered hydroxyapatite/titanium composite coatings. Biomaterials 24:4233-4238

97. Dorozhkin SV (2009) Calcium orthophosphates in nature, biology and medicine. Materials 2:399-498

98. Dorozhkin SV (2011) Calcium orthophosphates: occurrence, properties, biomineralization, pathological calcification and biomimetic applications. Biomatter 1:121-164

99. Dorozhkin SV (2012) Calcium orthophosphates: applications in nature, biology, and medicine. Pan Stanford, Singapore. p 850

100. Dorozhkina EI, Dorozhkin SV (2003) Structure and properties of the precipitates formed from condensed solutions of the revised simulated body fluid. J Biomed Mater Res A 67A:578-581

101. Dostálová T, Himmlová L, Jélinek M, Grivas C (2001) Osseointegration of loaded dental implant with KrF laser hydroxylapatite films on Ti6Al4V alloy by minipigs. J. Biomed. Optics 6:239-243

102. Duan K, Fan Y, Wang R (2003) Electrochemical deposition and patterning of calcium phosphate bioceramic coating. Ceram Transact 147:53-61

103. Ducheyne P, Healy KE (1988) The effect of plasma-sprayed calcium phosphate ceramic coatings on the metal ion release from porous titanium and cobalt-chromium alloys. J Biomed Mater Res 22:1137-1163

104. Ducheyne P, van Raemdonck W, Heughebaert JC, Heughebaert M (1986) Structural analysis of hydroxylapatite coatings on titanium. Biomaterials 7:97-103

105. Ducheyne P, Radin S, Heughebaert M, Heughebaert JC (1990) Calcium phosphate ceramic coatings on porous titanium: effect of structure and composition on electrophoretic deposition, vacuum sintering and in vitro dissolution. Biomaterials 11:244-254

106. Dyshlovenko S, Pateyron B, Pawlowski L, Murano D (2004) Numerical simulation of hydroxyapatite powder behaviour in plasma jet. Surf CoatTechnol 179:110-117 corrigendum: Surf. Coat. Technol. 2004, 187, 408–409

107. Eliaz N, Eliyahu M (2007) Electrochemical processes of nucleation and growth of hydroxyapatite on titanium supported by realtime electrochemical atomic force microscopy. J Biomed Mater Res A 80A:621-634

108. Elliott JC (1994) Structure and chemistry of the apatites and other calcium orthophosphates. In: Studies in inorganic chemistry. Elsevier, Amsterdam, Netherlands. p 38918.

109. Epinette JAMD, Geesink RGT (1995) Hydroxyapatite coated hip and knee arthroplasty. Elsevier, Amsterdam, Netherlands. p 394

110. Epinette JA, Manley MT (eds) (2004) Fifteen years of clinical experience with hydroxyapatite coatings in joint arthroplasty Springer, France. p 452

111. Erkmen ZE (1999) The effect of heat treatment on the morphology of D-gun sprayed hydroxyapatite coatings. J Biomed Mater Res (Appl Biomater) 48:861-868

112. Falguera V, Quintero JP, Jiménez A, Muñoz JA, Ibarz A (2011) Edible films and coatings: structures, active functions and trends in their use. Trends Food Sci Technol 22:292-303

113. Fauchais P (2004) Understanding plasma spraying. J Phys D: Appl Phys 37:R86-R108

114. Fauchais P, Vardelle A, Dussoubs B (2001) Quo vadis thermal spraying? J Thermal Spray Technol 10:44-66

115. Feddes B, Wolke JGC, Jansen JA, Vredenberg AM (2003) Radio frequency magnetron sputtering deposition of calcium phosphate coatings: Monte Carlo simulations of the deposition process and depositions through an aperture. J Appl Phys 93:662-670

116. Feddes B, Wolke JGC, Jansen JA, Vredenberg AM (2003) Radio frequency magnetron sputtering deposition of calcium phosphate coatings: the effect of resputtering on the coating composition. J Appl Phys 93:9503-9507

117. Feng B, Chen Y, Zhang XD (2002) Effect of water vapor treatment on apatite formation on precalcified titanium and bond strength of coatings to substrates. J Biomed Mater Res 59:12-17

118. Fernández-Pradas JM, Sardin G, Clèrics L, Serra P, Ferrater C, Morenza JL (1998) Deposition of hydroxyapatite thin films by excimer laser ablation. Thin Solid Films 317:393-396

119. Fernandez-Pradas JM, Cleries L, Sardin G, Morenza JL (1999) Hydroxyapatite coatings grown by pulsed laser deposition with a beam of 355 nm wavelength. J Mater Res 14:4715-4719

120. Fernandez-Pradas JM, Cleries L, Martinez E, Sardin G, Esteve J, Morenza JL (2000) Calcium phosphate coatings deposited by laser ablation at 355 nm under different substrate temperatures and water vapour pressures. Appl. Phys. A 71:37-42.

121. Fernandez-Pradas JM, Cleries L, Martinez E, Sardin G, Esteve J, Morenza JL (2001) Influence of thickness on the properties of hydroxyapatite coatings deposited by KrF laser ablation. Biomaterials 22:2171-2175

122. Filiaggi MJ, Coombs NA, Pilliar RM (1991) Characterization of the interface in plasma-sprayed HA coating/Ti–6Al–4V implant system. J Biomed Mater Res 25:1211-1229

123. Freidberg JP (2007) Plasma physics and fusion energy. Cambridge University Press, Cambridge, UK. p 692

124. Fujihara T, Tsukamoto M, Abe N, Miyake S, Ohji T, Akedo J (2004) Hydroxyapatite film formed by beam irradiation. Vacuum 73:629-633

125. Furlong RJ, Osborn JF (1991) Fixation of hip prostheses by hydroxyapatite ceramic coating. J. Bone Joint Surg. B 73:741-745

126. Gan L, Pilliar R (2004) Calcium phosphate sol–gel-derived thin films on porous surfaced implants for enhanced osteoconductivity. Part I: Synthesis and characterization. Biomaterials 25:5303-5312

127. Gan L, Wang J, Tache A, Valiquette N, Deporter D, Pilliar R (2004) Calcium phosphate sol–gel-derived thin films on porous-surfaced implants for enhanced osteoconductivity. Part II: short-term in vivo studies. Biomaterials 25:5313-5321

128. Gandhi R, Davey JR, Mahomed NN (2009) Hydroxyapatite coated femoral stems in primary total hip arthroplasty. A meta-analysis. J. Arthroplasty 24:38-42

129. Gärtner F, Stoltenhoff T, Schmidt T, Kreye H (2006) The cold spray process and its potential for industrial applications. J. Thermal Spray Technol. 15:223-232

130. Geesink RGT (1990) Hydroxyapatite-coated total hip prostheses; two-year clinical and roentgenographic results of 100 cases. Clin. Orthop. Rel. Res. 261:39-58

131. Geesink RGT (2002) Osteoconductive coating for total joint arthroplasty. Clin Orthop Rel Res 395:53-65

132. Geesink RGT, Hoefnagels NHM (1995) Six-year results of hydroxyapatite-coated total hip replacement. J Bone Jt Surg Br 77B:534-547

133. Geesink RGT, de Groot K, Klein CPAT (1987) Chemical implant fixation using hydroxyl-apatite coatings. The development of a human total hip prosthesis for chemical fixation to bone using hydroxyl-apatite coatings on titanium substrates. Clin. Orthop. Rel. Res. 225:147-170

134. Gill WD, Kay E (1965) Efficient low pressure sputtering in a large inverted magnetron suitable for film synthesis. Rev. Scientif. Instrum. 36:277-282

135. Gineste L, Gineste M, Ranz X, Ellefterion A, Guilhem A, Rouquet N, Frayssinet P (1999) Degradation of hydroxylapatite, fluorapatite, and fluorhydroxyapatite coatings of dental implants in dogs. J Biomed Mater Res 48:224-234

136. Gledhill HC, Turner IG, Doyle C (1999) Direct morphological comparison of vacuum plasma sprayed and detonation gun sprayed hydroxyapatite coatings. Biomaterials 20:315-322

137. Gledhill HC, Turner IG, Doyle C (2001) In vitro fatigue behavior of vacuum plasma and detonation gun sprayed hydroxyapatite coatings. Biomaterials 22:1233-1240

138. Gottlander M, Johansson CB, Wennerberg A, Albrektsson T, Radin S, Ducheyne P (1997) Bone tissue reactions to an electrophoretically applied calcium phosphate coating. Biomaterials 18:551-557

139. Gottlander M, Johansson CB, Albrektsson T (1997) Short- and long-term animal studies with a plasma-sprayed calcium phosphate-coated implant. Clin. Oral Implant. Res. 8:345-351

140. Granato R, Marin C, Suzuki M, Gil JN, Janal MN, Coelho PG (2009) Biomechanical and histomorphometric evaluation of a thin ion beam bioceramic deposition on plateau root form implants: an experimental study in dogs. J Biomed Mater Res B Appl Biomater 90B:396-403

141. Gross KA, Babovic M (2002) Influence of abrasion on the surface characteristics of thermally sprayed hydroxyapatite coatings. Biomaterials 23:4731-4737

142. Gross KA, Berndt CC (1994) In vitro testing of plasma-sprayed hydroxyapatite coatings. J Mater Sci Mater Med 5:219-224

143. Gross KA, Saber-Samandari S (2009) Revealing mechanical properties of a suspension plasma sprayed coating with nanoindentation. Surf CoatTechnol 203:2995-2999

144. Gross KA, Berndt CC, Goldschlag DD, Iacono VJ (1997) In vitro changes of hydroxyapatite coatings. Int J Oral Maxillofac Implants 12:589-597

145. Gross KA, Gross V, Berndt CC (1998) Thermal analysis of amorphous phases in hydroxyapatite coatings. J Am Ceram Soc 81:106-112

146. Gross KA, Berndt CC, Herman H (1998) Amorphous phase formation in plasma-sprayed hydroxyapatite coatings. J Biomed Mater Res 39:407-414

147. Gross KA, Chai CS, Kannangara GSK, Ben-Nissan B, Hanley L (1998) Thin hydroxyapatite coatings via sol–gel synthesis. J Mater Sci Mater Med 9:839-843

148. Gu Y, Meng G (1999) A model for ceramic membrane formation by dip-coating. J Eur Ceram Soc 19:1961-1966

149. Habibovic P, Barre're F, van Blitterswijk CA, de Groot K, Layrolle P (2002) Biomimetic hydroxyapatite coating on metal implants. J Am Ceram Soc 85:517-522.

150. Habibovic P, Li J, van der Valk CM, Meijer G, Layrolle P, van Blitterswijk CA, de Groot K (2005) Biological performance of uncoated and octacalcium phosphate-coated Ti6Al4V. Biomaterials 26:23-36

151. Haddow DB, James PF, van Noort R (1999) Sol–gel derived calcium phosphate coatings for biomedical applications. J Sol–gel Part Sci Technol 13:261-265

152. Hahn BD, Park DS, Choi JJ, Ryu J, Yoon WH, Kim KH, Park C, Kim HE (2009) Dense nanostructured hydroxyapatite coating on titanium by aerosol deposition. J Am Ceram Soc 92:683-687

153. Haman JD, Chittur KK, Crawmer DE, Lucas LC (1999) Analytical and mechanical testing of high velocity oxy-fuel thermal sprayed and plasma sprayed calcium phosphate coatings. J Biomed Mater Res (Appl Biomater) 48:856-860

154. Hamdi M, Ide-Ektessabi A (2003) Preparation of hydroxyapatite layer by ion beam assisted simultaneous vapor deposition. Surf CoatTechnol 163–164:362-367

155. Hamdi M, Toque JA, Ide-Ektessabi A (2010) Wear characteristics and adhesion behavior of calcium phosphate thin-films. Key Eng. Mater. 443:469-474

156. Han Y, Xu KW, Lu J, Wu Z (1999) The structural characteristics and mechanical behaviors of nonstoichiometric apatite coatings sintered in air atmosphere. J Biomed Mater Res 45:198-203

157. Han Y, Xu K, Lu J (1999) Morphology and composition of hydroxyapatite coatings prepared by hydrothermal treatment on electrodeposited brushite coatings. J Mater Sci Mater Med 10:243-248

158. Han Y, Fu T, Lu J, Xu K (2001) Characterization and stability of hydroxyapatite coatings prepared by an electrodeposition and alkaline-treatment process. J Biomed Mater Res 54:96-101

159. Han Y, Sun J, Huang X (2008) Formation mechanism of HA-based coatings by micro-arc oxidation. Electrochem. Comm. 10:510-513

160. Hanawa T, Ota M (1992) Characterization of surface film formed on titanium in electrolyte using XPS. Appl Surf Sci 55:269-276

161. Hasan S, Stokes J (2011) Design of experiment analysis of the Sulzer Metco DJ high velocity oxy-fuel coating of hydroxyapatite for orthopedic applications. J Thermal Spray Technol 20:186-194

162. Hashimoto Y, Kawashima M, Hatanaka R, Kusunoki M, Nishikawa H, Hontsu S, Nakamura M (2008) Cytocompatibility of calcium phosphate coatings deposited by an ArF pulsed laser. J Mater Sci Mater Med 19:327-333

163. Hayakawa T, Yoshinari M, Nemoto K, Wolke JGC, Jansen JA (2000) Effect of surface roughness and calcium phosphate coating on the implant/bone response. Clin. Oral Implant. Res. 11:296-304

164. Hayakawa T, Yoshinari M, Kiba H, Yamamoto H, Nemoto K, Jansen JA (2002) Trabecular bone response to surface roughened and calcium phosphate (Ca-P) coated titanium implants. Biomaterials 23:1025-1031

165. Heimann RB (2006) Thermal spraying of biomaterials. Surf CoatTechnol 201:2012-2019

166. Heimann RB (2009) Characterization of as-plasma-sprayed and incubated hydroxyapatite coatings with high resolution techniques. Mater Werkst 40:23-30

167. Heimann RB, Wirth R (2006) Formation and transformation of amorphous calcium phosphates on titanium alloy surfaces during atmospheric plasma spraying and their subsequent in vitro performance. Biomaterials 27:823-831

168. Heimann RB, Schürmann N, Müller RT (2004) In vitro and in vivo performance of Ti6Al4V implants with plasma-sprayed osteoconductive hydroxylapatite-bioinert titania bond coat "duplex" systems: an experimental study in sheep. J Mater Sci Mater Med 15:1045-1052

169. Hench LL (1991) Bioceramics: from concept to clinic. J Am Ceram Soc 74:1487-1510

170. Herman H (1988) Plasma-sprayed coatings. Sci Am 9:112-117

171. Herø H, Wie H, Jorgensen RB, Ruyter IE (1994) Hydroxyapatite coatings on Ti produced by hot isostatic pressing. J Biomed Mater Res 28:343-348

172. Hontsu S, Matsumoto T, Ishii J, Nakamori M, Tabata H, Kawai T (1997) Electrical properties of hydroxyapatite thin films grown by pulsed laser deposition. Thin Solid Films 295:214-217

173. Hou X, Choy KL, Leach SE (2007) Processing and in vitro behavior of hydroxyapatite coatings prepared by electrostatic spray assisted vapor deposition method. J Biomed Mater Res A 83A:683-691

174. Huang Y, Qu Y, Yang B, Li W, Zhang B, Zhang X (2009) In vivo biological responses of plasma sprayed hydroxyapatite coatings with an electric polarized treatment in alkaline solution. Mater Sci Eng C 29:2411-2416

175. Huang Y, Song L, Liu X, Xiao Y, Wu Y, Chen J, Wu F, Gu Z (2010) Hydroxyapatite coatings deposited by liquid precursor plasma spraying: controlled dense and porous microstructures and osteoblastic cell responses. Biofabrication 2:045003

176. Hughes JM, Kohn M, Rakovan J (eds) (2002) Reviews in mineralogy and geochemistry. Phosphates: geochemical, geobiological and materials importance. vol. 48 Mineralogical Society of America, Washington, DC, USA. p 742

177. Hulshoff JEG, Jansen JA (1997) Initial interfacial healing events around calcium phosphate (Ca-P) coated oral implants. Clin. Oral Implant. Res. 8:393-400

178. Hulshoff JEG, van Dijk K, van der Waerden JPCM, Kalk W, Jansen JA (1996) A histological and histomorphometrical evaluation of screw-type calciumphosphate (Ca-P) coated implants; an in vivo experiment in maxillary cancellous bone of goats. J Mater Sci Mater Med 7:603-609

179. Hulshoff JEG, van Dijk K, van Der Waerden JPCM, Wolke JGC, Kalk W, Jansen JA (1996) Evaluation of plasma-spray and magnetron-sputter Ca-P-coated implants: an in vivo experiment using rabbits. J Biomed Mater Res 31:329-337

180. Hulshoff JEG, Hayakawa T, van Dijk K, Leijdekkers-Govers AFM, van der Waerden JPCM, Jansen JA (1997) Mechanical and histologic evaluation of Ca-P plasma-spray and magnetron sputter-coated implants in trabecular bone of the goat. J Biomed Mater Res 36:75-83

181. Iafisco M, Bosco R, Leeuwenburgh SCG, van den Beucken JJJP, Jansen JA, Prat M, Roveri N (2012) Electrostatic spray deposition of biomimetic nanocrystalline apatite coatings onto titanium. Adv Eng Mater 14:B13-B20

182. Ievlev VM, Domashevskaya EP, Putlyaev VI, Tret'yakov YD, Barinov SM, Belonogov EK, Kostyuchenko AV, Petrzhik MI, Kiryukhantsev-Korneev FV (2008) Structure, elemental composition, and mechanical properties of films prepared by radio-frequency magnetron sputtering of hydroxyapatite. Glass Phys Chem 34:608-616

183. Iezzi G, Scarano A, Petrone G, Piattelli A (2007) Two human hydroxyapatite-coated dental implants retrieved after a 14-year loading period: a histologic and histomorphometric case report. J Periodontol 78:940-947

184. ISO (1996) Implants for surgery: coating for hydroxyapatite ceramics. ISO, Geneva. pp 1-8

185. ISO (2000) ISO 13779–2 Implants for surgery – Hydroxyapatite – Part 2: Coatings.of.hydroxyapatite.http://www.iso.org/iso/iso_catalogue/catalogue_tc/catalogue_detail.htm?csnumber=26841.

186. ISO (2005) ISO 20502. Fine ceramics (advanced ceramics, advanced technical ceramics) – determination of adhesion of ceramic coatings by scratch testing.http://www.iso.org/iso/iso_catalogue/catalogue_tc/catalogue_detail.htm?csnumber=34189.

187. ISO (2008) ISO 13779–2 Implants for surgery – Hydroxyapatite – Part 2: Coatingsof.hydroxyapatite.http://www.iso.org/iso/iso_catalogue/catalogue_tc/catalogue_detail.htm?csnumber=43827

188. Jansen JA, van der Waerden JPCM, Wolke JGC, de Groot K (1991) Histologic evaluation of the osseous adaptation to titanium hydroxyapatite-coated implants. J Biomed Mater Res 25:973-989

189. Jansen JA, Wolke JGC, Swann S, van der Waerden JPCM, de Groot K (1993) Application of magnetron sputtering for producing ceramic coatings on implant materials. Clin. Oral Impl. Res. 4:28-34

190. Jaworski R, Pierlot C, Pawlowski L, Bigan M, Martel M (2009) Design of the synthesis of fine HA powder for suspension plasma spraying. Surf CoatTechnol 203:2092-2097

191. Jedynski M, Hoffman J, Mroz W, Szymanski Z (2008) Plasma plume induced during ArF laser ablation of hydroxyapatite. Appl Surf Sci 255:2230-2236

192. Ji H, Marquis PM (1993) Effect of heat treatment on the microstructure of plasma-sprayed hydroxyapatite coating. Biomaterials 14:64-68

193. Jiang G, Shi D (1998) Coating of hydroxyapatite on highly porous Al2O3 substrate for bone substitutes. J. Biomed. Mater. Res. (Appl. Biomater.) 43:77-81

194. Jiang W, Sun L, Nyandoto G, Malshe AP (2008) Electrostatic spray deposition of nanostructured hydroxyapatite coating for biomedical applications. J. Manufact. Sci. Eng. Transact. ASME 130:0210011-0210017

195. Johnson S, Haluska M, Narayan RJ, Snyder RL (2006) In situ annealing of hydroxyapatite thin films. Mater Sci Eng C 26:1312-1316

196. Jung YC, Han CH, Lee IS, Kim HE (2001) Effects of ion beam-assisted deposition of hydroxyapatite on the osseointegration of endosseous implants in rabbit tibiae. Int J Oral Maxillofac Implants 16:809-818

197. Junker R, Manders PJD, Wolke J, Borisov Y, Jansen JA (2010) Bone-supportive behavior of microplasma-sprayed CaP-coated implants: mechanical and histological outcome in the goat. Clin. Oral Implant. Res. 21:189-200

198. Kameyama T (1999) Hybrid bioceramics with metals and polymers for better biomaterials. Bull Mater Sci 22:641-646

199. Katayama H, Katto M, Nakayama T (2009) Laser-assisted laser ablation method for high-quality hydroxyapatite coating onto titanium substrate. Surf CoatTechnol 204:135-140

200. Kelly PJ, Arnell RD (2000) Magnetron sputtering: a review of recent developments and applications. Vacuum 56:159-172

201. Khor KA, Cheang P (1993) Characterization of plasma sprayed hydroxyapatite powders and coatings. In: Berndt CC, Bernecki TF (eds) Thermal spray coatings: research, design and applications, ASM International, Materials Park, Ohio, USA. pp 347-352

202. Khor KA, Li H, Cheang P (2003) Characterization of the bone-like apatite precipitated on high velocity oxy-fuel (HVOF) sprayed calcium phosphate deposits. Biomaterials 24:769-775

203. Khor KA, Li H, Cheang P (2003) Processing-microstructure-property relations in HVOF sprayed calcium phosphate based bioceramic coatings. Biomaterials 24:2233-2243

204. Khor KA, Li H, Cheang P (2004) Significance of melt-fraction in HVOF sprayed hydroxyapatite particles, splats and coatings. Biomaterials 25:1177-1186

205. Kim HM, Miyaji F, Kokubo T, Nakamura T (1997) Bonding strength of bonelike apatite layer to Ti metal substrate. J Biomed Mater Res 38:121-127

206. Kim TN, Feng QL, Luo ZS, Cui FZ, Kim JO (1998) Highly adhesive hydroxyapatite coatings on alumina substrates prepared by ion-beam assisted deposition. Surf Coat Technol 99:20-23

207. Kim HM, Kishimoto K, Miyaji F, Kokubo T, Yao T, Suetsugu Y, Tanaka J, Nakamura T (2000) Composition and structure of apatite formed on organic polymer in simulated body fluid with a high content of carbonate ion. J Mater Sci Mater Med 11:421-426

208. Kim HW, Knowles JC, Salih V, Kim HE (2004) Hydroxyapatite and fluor-hydroxyapatite layered film on titanium processed by a sol–gel route for hard-tissue implants. J Biomed Mater Res B Appl Biomater 71B:66-76

209. Kim HW, Koh YH, Li LH, Lee S, Kim HE (2004) Hydroxyapatite coating on titanium substrate with titania buffer layer processed by sol–gel method. Biomaterials 25:2533-2538

210. Kim H, Vohra YK, Louis PJ, Lacefield WR, Lemons JE, Camata RP (2005) Biphasic and preferentially oriented microcrystalline calcium phosphate coatings: in-vitro and in-vivo studies. Key Eng. Mater. 284–286:207-210

211. Kim H, Camata RP, Vohra YK, Lacefield WR (2005) Control of phase composition in hydroxyapatite/tetracalcium phosphate biphasic thin coatings for biomedical applications. J Mater Sci Mater Med 16:961-966

212. Kim H, Camata RP, Lee S, Rohrer GS, Rollett AD, Vohra YK (2007) Crystallographic texture in pulsed laser deposited hydroxyapatite bioceramic coatings. Acta Mater 55:131-139

213. Kim BH, Jeong JH, Jeon YS, Jeon KO, Hwang KS (2007) Hydroxyapatite layers prepared by sol–gel assisted electrostatic spray deposition. Ceram Int 33:119-122

214. Kim H, Camata RP, Chowdhury S, Vohra YK (2010) In vitro dissolution and mechanical behavior of c-axis preferentially oriented hydroxyapatite thin films fabricated by pulsed laser deposition. Acta Biomater 6:3234-3241

215. Klein CPAT, Patka P, van der Lubbe HBM, Wolcke JGC, de Groot K (1991) Plasma-sprayed coatings of tetracalcium phosphate, hydroxylapatite, and α-TCP on titanium alloy: an interface study. J Biomed Mater Res 25:53-65

216. Klein CPAT, Patka P, Wolke JGC, de Blieck-Hogervorst JMA, de Groot K (1994) Long-term in vivo study of plasma-sprayed coatings on titanium alloys of tetracalcium phosphate, hydroxyapatite and α-tricalcium phosphate. Biomaterials 15:146-150

217. Klein CPAT, Wolke JGC, de Blieck-Hogervorst JMA, de Groot K (1994) Calcium phosphate plasma-sprayed coatings and their stability: an in vivo study. J Biomed Mater Res 28:909-917

218. Kobayashi T, Itoh S, Nakamura S, Nakamura M, Shinomiya K, Yamashita K (2007) Enhanced bone bonding of hydroxyapatite-coated titanium implants by electrical polarization. J Biomed Mater Res A 82A:145-151

219. Koch B, Wolke JGC, de Groot K (1990) X-ray diffraction studies on plasma-sprayed calcium phosphate-coated implants. J Biomed Mater Res 24:655-667

220. Koch CF, Johnson S, Kumar D, Jelinek M, Chrisey DB, Doraiswamy A, Jin C, Narayan RJ, Mihailescu IN (2007) Pulsed laser deposition of hydroxyapatite thin films. Mater Sci Eng C 27:484-494

221. Kokubo T (ed) (2008) Bioceramics and their clinical applications Woodhead Publishing, Abington, Cambridge, UK. p 784

222. Kokubo T, Yamaguchi S (2011) Bioactive layer formation on metals and polymers. In: Ducheyne P, Healy K, Hutmacher DW, Grainger DW, Kirkpatrick CJ (eds) Comprehensive biomaterials, vol. 1, Elsevier, Amsterdam, Netherlands. pp 231-244

223. Kokubo T, Miyaji F, Kim HM (1996) Spontaneous formation of bonelike apatite layer on chemically treated titanium metals. J Am Ceram Soc 79:1127-1129

224. Kokubo T, Kim HM, Kawashita M (2003) Novel bioactive materials with different mechanical properties. Biomaterials 24:2161-2175

225. Kumar M, Dasarathy H, Riley C (1999) Electrodeposition of brushite coatings and their transformation to hydroxyapatite in aqueous solutions. J Biomed Mater Res 45:302-310

226. Kummer FJ, Jaffe WL (1992) Stability of a cyclically loaded hydroxylapatite coating: effect of substrate material, surface, preparation and testing environment. J. Appl. Mater. 3:211-215

227. Kuo MC, Yen SK (2002) The process of electrochemical deposited hydroxyapatite coatings on biomedical titanium at room temperature. Mater Sci Eng C 20:153-160

228. Kurella A, Dahotre NB (2006) A multi-textured calcium phosphate coating for hard tissue via laser surface engineering. JOM 58:64-66

229. Kuroda K, Miyashita Y, Ichino R, Okido M, Takai O (2002) Preparation of calcium phosphate coatings on titanium using the thermal substrate method and their in vitro evaluation. Mater Transact 43:3015-3019

230. Kuroda K, Ichino R, Okido M, Takai O (2002) Effects of ion concentration and pH on hydroxyapatite deposition from aqueous solution onto titanium by the thermal substrate method. J Biomed Mater Res 61:354-359

231. Kuroda K, Miyashita Y, Ichino R, Okido M (2003) Hydroxyapatite coating on titanium by thermal substrate method in an aqueous solution and its behavior in SBF. Mater Sci Forum 426–432:3189-3194

232. Kuroda K, Nakamoto S, Ichino R, Okido M, Pilliar RM (2005) Hydroxyapatite coatings on a 3D porous surface using thermal substrate method. Mater. Transact. 46:1633-1635

233. Kuroda K, Nakamoto S, Miyashita Y, Ichino R, Okido M (2007) Osteoinductivity of hydroxyapatite films with different surface morphologies coated by the thermal substrate method in aqueous solutions. J. Jpn. Inst. Metals 71:342-345

234. Kweh SWK, Khor KA, Cheang P (2000) Plasma-sprayed hydroxyapatite (HA) coatings with flame-spheroidized feedstock: microstructure and mechanical properties. Biomaterials 21:1223-1234

235. Lacefield WR (1988) Hydroxyapatite coatings. Ann. New York Acad. Sci. 523:72-80

236. Łatka L, Pawlowski L, Chicot D, Pierlot C, Petit F (2010) Mechanical properties of suspension plasma sprayed hydroxyapatite coatings submitted to simulated body fluid. Surf CoatTechnol 205:954-960

237. Lavos-Valereto IC, Wolynec S, Deboni MCZ, Knig B Jr (2001) In vitro and in vivo biocompatibility testing of Ti-6Al-7Nb alloy with and without plasma-sprayed hydroxyapatite coating. J Biomed Mater Res 58:727-733

238. Layrolle P (2011) Calcium phosphate coatings. In: Ducheyne P, Healy K, Hutmacher DW, Grainger DW, Kirkpatrick CJ (eds) Comprehensive biomaterials, vol. 1, Elsevier, Amsterdam, Netherlands. pp 223-229 Vol. 1.

239. Lazarinis S, Krärholm J, Hailer NP (2010) Increased risk of revision of acetabular cups coated with hydroxyapatite: a Swedish Hip Arthroplasty Register study involving 8,043 total hip replacements. Acta Orthop 81:53-59

240. Lazarinis S, Krrholm J, Hailer NP (2011) Effects of hydroxyapatite coating on survival of an uncemented femoral stem. Acta Orthop 82:399-404

241. Lee JJ, Rouhfar L, Beirne OR (2000) Survival of hydroxyapatite-coated implants: a meta-analytic review. J Oral Maxillofac Surg 58:1372-1379

242. Lee IS, Whang CN, Lee GH, Cui FZ, Ito A (2003) Effects of ion beam assist on the formation of calcium phosphate film. Nucl Instr Methods Phys Res B 206:522-526

243. Lee YP, Wang CK, Huang TH, Chen CC, Kao CT, Ding SJ (2005) In vitro characterization of post heat-treated plasma-sprayed hydroxyapatite coatings. Surf CoatTechnol 197:367-374

244. Lee EJ, Lee SH, Kim HW, Kong YM, Kim HE (2005) Fluoridated apatite coatings on titanium obtained by electron-beam deposition. Biomaterials 26:3843-3851

245. Lee IS, Zhao B, Lee GH, Choi SH, Chung SM (2007) Industrial application of ion beam assisted deposition on medical implants. Surf CoatTechnol 201:5132-5137

246. Lee WH, Kim YH, Oh NH, Cheon YW, Cho YJ, Lee CM, Kim KB, Lee NS (2007) A study of hydroxyapatite coating on porous Ti compact by electrostatic spray deposition. Diffusion and Defect Data B. Solid State Phenomena 124–126:1789-1792

247. Lee JH, Kim SG, Lim SC (2011) Histomorphometric study of bone reactions with different hydroxyapatite coating thickness on dental implants in dogs. Thin Solid Films 519:4618-4622

248. Leeuwenburgh S, Wolke J, Schoonman J, Jansen J (2003) Electrostatic spray deposition (ESD) of calcium phosphate coatings. J Biomed Mater Res A 66A:330-334

249. Leeuwenburgh SC, Wolke JG, Schoonman J, Jansen JA (2004) Influence of precursor solution parameters on chemical properties of calcium phosphate coatings prepared using electrostatic spray deposition (ESD). Biomaterials 25:641-649

250. Leeuwenburgh SCG, Wolke JGC, Schoonman J, Jansen JA (2005) Influence of deposition parameters on morphological properties of biomedical calcium phosphate coatings prepared using electrostatic spray deposition. Thin Solid Films 472:105-113

251. Leeuwenburgh S, Wolke J, Schoonman J, Jansen JA (2005) Influence of deposition parameters on chemical properties of calcium phosphate coatings prepared by using electrostatic spray deposition. J Biomed Mater Res A 74A:275-284

252. Leeuwenburgh SCG, Wolke JGC, Schoonman J, Jansen JA (2006) Deposition of calcium phosphate coatings with defined chemical properties using the electrostatic spray deposition technique. J Eur Ceram Soc 26:487-493

253. Leeuwenburgh SCG, Heine MC, Wolke JGC, Pratsinis SE, Schoonman J, Jansen JA (2006) Morphology of calcium phosphate coatings for biomedical applications deposited using electrostatic spray deposition. Thin Solid Films 503:69-78

254. LeGeros RZ (1991) Calcium phosphates in oral biology and medicine. In: Myers HM (ed) Monographs in oral science, vol. 15, Karger, Basel, Switzerland. p 20115.

255. LeGeros RZ (2008) Calcium phosphate-based osteoinductive materials. Chem. Rev. 108:4742-4753

256. Leitão E, Barbosa MA, de Groot K (1995) In vitro calcification of orthopaedic implant materials. J Mater Sci Mater Med 6:849-852

257. León B, Jansen JA (eds) (2009) Thin calcium phosphate coatings for medical implants Springer, New York, USA. p 326

258. Li P (2003) Biomimetic nano-apatite coating capable of promoting bone ingrowth. J Biomed Mater Res A 66A:79-85

259. Li P, Ohtsuki C, Kokubo T, Nakanishi K, Soga N (1992) Apatite formation induced by silica gel in a simulated body fluid. J Am Ceram Soc 75:2094-2097

260. Li P, Ohtsuki C, Kokubo T, Nakanishi K, Soga N, de Groot K (1994) The role of hydrated silica, titania, and alumina in inducing apatite on implants. J Biomed Mater Res 28:7-15

261. Li T, Lee J, Kobayashi T, Aoki H (1996) Hydroxyapatite coating by dipping method, and bone bonding strength. J Mater Sci Mater Med 7:355-357

262. Li H, Khor KA, Cheang P (2000) Effect of the powders' melting state on the properties of HVOF sprayed hydroxyapatite coatings. Mater Sci Eng, A 293:71-80

263. Li H, Khor KA, Cheang P (2002) Properties of heat-treated calcium phosphate coatings deposited by high-velocity oxy-fuel (HVOF) spray. Biomaterials 23:2105-2112

264. Li F, Feng QL, Cui FZ, Li HD, Schubert H (2002) A simple biomimetic method for calcium phosphate coating. Surf Coat Technol 154:88-93

265. Liang F, Zhou L, Wang K (2003) Apatite formation on porous titanium by alkali and heat-treatment. Surf Coat Technol 165:133-139

266. Lim YM, Kim BH, Jeon YS, Jeon KO, Hwang KS (2005) Calcium phosphate films deposited by electrostatic spray deposition and an evaluation of their bioactivity. J. Ceram. Process. Res. 6:255-258

267. Lin S, LeGeros RZ, LeGeros JP (2003) Adherent octacalciumphosphate coating on titanium alloy using modulated electrochemical deposition method. J Biomed Mater Res A 66A:819-828

268. Liu DM, Troczynski T, Hakimi D (2002) Effect of hydrolysis on the phase evolution of water-based sol–gel hydroxyapatite and its application to bioactive coatings. J Mater Sci Mater Med 13:657-665

269. Liu D, Yang Q, Troczynski T (2002) Sol–gel hydroxyapatite coatings on stainless steel substrates. Biomaterials 23:691-698

270. Liu X, Chu PK, Ding C (2004) Surface modification of titanium, titanium alloys, and related materials for biomedical applications. Mater. Sci. Eng. R 47:49-121

271. Liu F, Song Y, Wang F, Shimizu T, Igarashi K, Zhao L (2005) Formation characterization of hydroxyapatite on titanium by microarc oxidation and hydrothermal treatment. J Biosci Bioeng 100:100-104

272. Lo WJ, Grant DM, Ball MD, Welsh BS, Howdle SM, Antonov EN, Bagratashvili VN, Popov VK (2000) Physical, chemical, and biological characterization of pulsed laser deposited and plasma sputtered hydroxyapatite thin films on titanium alloy. J Biomed Mater Res 50:536-545

273. Long J, Sim L, Xu S, Ostrikov K (2007) Reactive plasma-aided RF sputtering deposition of hydroxyapatite bio-implant coatings. Chem. Vapor Deposition 13:299-306

274. Lopez-Heredia MA, Weiss P, Layrolle P (2007) An electrodeposition method of calcium phosphate coatings on titanium alloy. J Mater Sci Mater Med 18:381-390

275. Lu X, Zhao Z, Leng Y (2005) Calcium phosphate crystal growth under controlled atmosphere in electrochemical deposition. J Cryst Growth 284:506-516

276. Lü X, Lin X, Guan T, Gao B, Huang W (2011) Effect of the mass ratio of CaCO3 to CaHPO4· 2H2O on in situ synthesis of hydroxyapatite coating by laser cladding. Rare Metal Mater. Eng. 40:22-27

277. Lü X, Lin X, Cao Y, Hu J, Gao B, Huang W (2011) Effects of processing parameters and heat treatment on phase structure of

the hydroxyapatite coating on pure Ti surface by laser cladding in-situ synthesis. Rare Metal Mater. Eng. 40:714-717

278. Luo ZS, Cui FZ, Li WZ (1999) Low-temperature crystallization of calcium phosphate coatings synthesized by ion beam-assisted deposition. J Biomed Mater Res 46:80-86

279. Luo ZS, Cui FZ, Feng QL, Li HD, Zhu XD, Spector M (2000) In vitro and in vivo evaluation of degradability of hydroxyapatite coatings synthesized by ion beam-assisted deposition. Surf CoatTechnol 131:192-195

280. Lusquiños F, Pou J, Arias JL, Boutinguiza M, Léon B, Pérez-Amor M, Driessens FCM, Merry JC, Gibson I, Best S, Bonfield W (2001) Production of calcium phosphate coatings on Ti6Al4V obtained by Nd: yttrium-aluminum-garnet laser cladding. J Appl Phys 90:4231-4236

281. Lusquiños F, de Carlos A, Pou J, Arias JL, Boutinguiza M, León B, Pérez-Amor M, Driessens FCM, Hing K, Gibson I, Best S, Bonfield W (2003) Calcium phosphate coatings obtained by Nd: YAG laser cladding: physicochemical and biologic properties. J Biomed Mater Res A 64A:630-637

282. Lusquiños F, Pou J, Boutinguiza M, Quintero F, Soto R, León B, Pérez-Amor M (2005) Main characteristics of calcium phosphate coatings obtained by laser cladding. Appl Surf Sci 247:486-492

283. Lv X, Lin X, Hu J, Gao B, Huang W (2012) Phase evolution in calcium phosphate coatings obtained by in situ laser cladding. Mater Sci Eng C 32:872-877

284. Ma J, Liang CH, Kong LB, Wang C (2003) Colloidal characterization and electrophoretic deposition of hydroxyapatite on titanium substrate. J Mater Sci Mater Med 14:797-801

285. Ma J, Wang C, Peng KW (2003) Electrophoretic deposition of porous hydroxyapatite scaffold. Biomaterials 24:3505-3510

286. Maistrelli GL, Mahomed N, Fornasier V, Antonelli L, Li Y, Binnington A (1993) Functional osseointegration of hydroxyapatite-coated implants in a weight bearing canine model. J Arthroplasty 8:549-554

287. Man HC, Chiu KY, Cheng FT, Wong KH (2009) Adhesion study of pulsed laser deposited hydroxyapatite coating on laser surface nitrided titanium. Thin Solid Films 517:5496-5501

288. MaNally SA, Shepperd HAN, Mann CV, Walczak JP (2000) The results at nine to twelve years of the use of a hydroxyapatite-coated femoral stem. J Bone Jt Surg Br 82B:378-382

289. Manders PJD, Wolke JGC, Jansen JA (2006) Bone response adjacent to calcium phosphate electrostatic spray deposition coated implants: an experimental study in goats. Clin. Oral Implant. Res. 17:548-553

290. Manso M, Jimenez C, Morant C, Herrero P, Martinez-Duart JM (2000) Electrodeposiiton of hydroxyapatite coatings in basic conditions. Biomaterials 21:1755-1761

291. Manso M, Langletm M, Jimenezm C, Martinez-Duart JM (2002) Microstructural study of aerosol-gel derived hydroxyapatite coatings. Biomol Eng 19:63-66

292. Manso M, Martínez-Duart JM, Langlet M, Jiménez C, Herrero P, Millon E (2002) Aerosol-gel-derived microcrystalline hydroxyapatite coatings. J Mater Res 17:1482-1489

293. Manso M, Ogueta S, Herrero-Fernández P, Vázquez L, Langlet M, García-Ruiz JP (2002) Biological evaluation of aerosol-gel-derived hydroxyapatite coatings with human mesenchymal stem cells. Biomaterials 23:3985-3990

294. Manso-Silván M, Langlet M, Jiménez C, Fernández M, Martínez-Duart JM (2003) Calcium phosphate coatings prepared by aerosol-gel. J Eur Ceram Soc 23:243-246

295. Massaro C, Baker MA, Cosentino F, Ramires PA, Klose S, Milella E (2001) Surface and biological evaluation of hydroxyapatite-based coatings on titanium deposited by different techniques. J Biomed Mater Res 58:651-657

296. Matsumine A, Myoui A, Kusuzaki K, Araki N, Seto M, Yoshikawa H, Uchida A (2004) Calcium hydroxyapatite ceramic implants in bone tumor surgery. A long-term follow-up study. J. Bone Joint Surg. B 86:719-725

297. Mayor B, Arias J, Chiussi S, Garcia F, Pou J, Fong BL, Pérez-Amor M (1998) Calcium phosphate coatings grown at different substrate temperatures by pulsed ArF-laser deposition. Thin Solid Films 317:363-366

298. McHugh TH (2000) Protein-lipid interactions in edible films and coatings. Nahrung 44:148-151

299. McPherson R, Gane N, Bastow TJ (1995) Structural characterization of plasma-sprayed hydroxylapatite coatings. J Mater Sci Mater Med 6:327-334

300. Mello A, Hong Z, Rossi AM, Luan L, Farina M, Querido W, Eon J, Terra J, Balasundaram G, Webster T, Feinerman A, Ellis DE, Ketterson JB, Ferreira CL (2007) Osteoblast proliferation on hydroxyapatite thin coatings produced by right angle magnetron sputtering. Biomed Mater 2:67-77

301. Meng X, Kwon TY, Kim KH (2006) Different morphology of hydroxyapatite coatings on titanium by electrophoretic deposition. Key Eng. Mater. 309–311:639-642

302. Meng X, Kwon TY, Kim KH (2008) Hydroxyapatite coating by electrophoretic deposition at dynamic voltage. Dent Mater J 27:666-671

303. Mennicke U, Salditt T (2002) Preparation of solid-supported lipid bilayers by spin-coating. Langmuir 18:8172-8177

304. Metikoš-Huković M, Tkalacec E, Kwokal A, Piljac J (2003) An in vitro study of Ti and Ti-alloys coated with sol–gel derived hydroxyapatite coatings. Surf CoatTechnol 165:40-50

305. Miroiu FM, Socol G, Visan A, Stefan N, Craciun D, Craciun V, Dorcioman G, Mihailescu IN, Sima LE, Petrescu SM, Andronie A, Stamatin I, Moga S, Ducu C (2010) Composite biocompatible hydroxyapatite-silk fibroin coatings for medical implants obtained by matrix assisted pulsed laser evaporation. Mater. Sci. Eng. B 169:151-158

306. Mistry S, Kundu D, Datta S, Basu D (2011) Comparison of bioactive glass coated and hydroxyapatite coated titanium dental implants in the human jaw bone. Australian Dent. J. 56:68-75

307. Miyaji F, Kim HM, Handa S, Kokubo T, Nakamura T (1999) Bonelike apatite coating on organic polymers: novel nucleation process using sodium silicate solution. Biomaterials 20:913-919

308. Mohammadi S, Esposito M, Hall J, Emanuelsson L, Krozer A, Thomsen P (2003) Short-term bone response to titanium implants coated with thin radiofrequent magnetron-sputtered hydroxyapatite in rabbits. Clin. Implant Dent. Rel. Res. 5:241-253

309. Mohammadi S, Esposito M, Hall J, Emanuelsson L, Krozer A, Thomsen P (2004) Long-term bone response to titanium implants coated with thin radiofrequent magnetron-sputtered hydroxyapatite in rabbits. Int J Oral Maxillofac Implants 19:498-509

310. Mondragón-Cortez P, Vargas-Gutiérrez G (2004) Electrophoretic deposition of hydroxyapatite submicron particles at high voltages. Mater Lett 58:1336-1339

311. Montenero A, Gnappi G, Ferrari F, Cesari M, Salvioli E, Mattogno L, Kaciulis S, Fini M (2000) Sol–gel derived hydroxyapatite coatings on titanium substrate. J Mater Sci 35:2791-2797

312. Morks MF, Kobayashi A (2007) Effect of gun current on the microstructure and crystallinity of plasma sprayed hydroxyapatite coatings. Appl Surf Sci 253:7136-7142

313. Morks MF, Kobayashi A, Fahim NF (2007) Abrasive wear behavior of sprayed hydroxyapitite coatings by gas tunnel type plasma spraying. Wear 262:204-209

314. Morris RE (ed) (2011) The sol–gel process: uniformity, polymers and applications Nova Science, Hauppauge, NY, USA. p 887

315. Muirhead-Allwood SK, Sandiford N, Skinner JA, Hua J, Kabir C, Walker PS (2010) Uncemented custom computer-assisted design and manufacture of hydroxyapatite-coated femoral components: survival at 10 to 17 years. J. Bone Joint Surg. B 92:1079-1084

316. Mukherjee DP, Dorairaj NR, Mills DK, Graham D, Krauser JT (2000) Fatigue properties of hydroxyapatite-coated dental implants after exposure to a periodontal pathogen. J Biomed Mater Res 53:467-474

317. Nanci A, Wuest JD, Peru L, Brunet P, Sharma V, Zalzal S, McKee MD (1998) Chemical modification of titanium surfaces for covalent attachment of biological molecules. J Biomed Mater Res 40:324-335

318. Narayanan R, Dutta S, Seshadri SK (2006) Hydroxy apatite coatings on Ti-6Al-4 V from seashell. Surf Coat Technol 200:4720-4730

319. Narayanan R, Seshadri SK, Kwon TY, Kim KH (2007) Electrochemical nano-grained calcium phosphate coatings on Ti-6Al-4 V for biomaterial applications. Scripta Mater. 56:229-232

320. Narayanan R, Kim SY, Kwon TY, Kim KH (2008) Nanocrystalline hydroxyapatite coatings from ultrasonated electrolyte: preparation, characterization and osteoblast responses. J Biomed Mater Res A 87A:1053-1060

321. Narayanan R, Kwon TY, Kim KH (2008) Direct nanocrystalline hydroxyapatite formation on titanium from ultrasonated electrochemical bath at physiological pH. Mater Sci Eng C 28:1265-1270

322. Narayanan R, Kwon TY, Kim KH (2008) Preparation and characteristics of nano-grained calcium phosphate coatings on titanium from ultrasonated bath at acidic pH. J Biomed Mater Res B Appl Biomater 85B:231-239

323. Narayanan R, Kim KH, Rautray TR (2010) Surface modification of titanium for biomaterial applications. Nova Science, Hauppauge, NY, USA. p 352

324. Negroiu G, Piticescu RM, Chitanu GC, Mihailescu IN, Zdrentu L, Miroiu M (2008) Biocompatibility evaluation of a novel hydroxyapatite-polymer coating for medical implants (in vitro tests). J Mater Sci Mater Med 19:1537-1544

325. Nelea V, Ristoscu C, Chiritescu C, Ghica C, Mihailescu IN, Pelletier H, Mille P, Cornet A (2000) Pulsed laser deposition of hydroxyapatite thin films on Ti-5Al-2.5Fe substrates with and without buffer layers. Appl Surf Sci 168:127-131

326. Nelea V, Pelletier H, Iliescu M, Werckmann J, Craciun V, Mihailescu IN, Ristoscu C, Ghica C (2002) Calcium phosphate thin film processing by pulsed laser deposition and in situ assisted ultraviolet pulsed laser deposition. J Mater Sci Mater Med 13:1167-1173

327. Nelea V, Morosanu C, Iliescu M, Mihailescu IN (2003) Microstructure and mechanical properties of hydroxyapatite thin films grown by RF magnetron sputtering. Surf CoatTechnol 173:315-322

328. Nelea V, Morosanu C, Iliescu M, Mihailescu IN (2004) Hydroxyapatite thin films grown by pulsed laser deposition and radio-frequency magnetron sputtering: comparative study. Appl Surf Sci 228:346-356

329. Nelea V, Morosanu C, Bercu M, Mihailescu IN (2007) Interfacial titanium oxide between hydroxyapatite and TiAlFe substrate. J Mater Sci Mater Med 18:2347-2354

330. Nguyen HQ, Deporter DA, Pilliar RM, Valiquette N, Yakubovich R (2004) The effect of sol–gel-formed calcium phosphate coatings on bone ingrowth and osteoconductivity of porous-surfaced Ti alloy implants. Biomaterials 25:865-876

331. Nie X, Leyland A, Matthews A (2000) Deposition of layered bioceramic hydroxyapatite/TiO2 coatings on titanium alloys using a hybrid technique of micro-arc oxidation and electrophoresis. Surf Coat Technol 125:407-414

332. Nie X, Leyland A, Matthews A, Jiang JC, Meletis EI (2001) Effects of solution pH and electrical parameters on hydroxyapatite coatings deposited by a plasma-assisted electrophoresis technique. J Biomed Mater Res 57:612-618

333. Nijhuis AWG, Leeuwenburgh SCG, Jansen JA (2010) Wet-chemical deposition of functional coatings for bone implantology. Macromol Biosci 10:1316-1329

334. Nimb L, Gotfredsen K, Steen JJ (1993) Mechanical failure of hydroxyapatite-coated titanium and cobalt-chromium-molybdenum alloyimplants. An animal study. Acta Orthop. Belg. 59:333-338

335. Ntsoane TP, Topic M, Bucher R (2011) Near-surface in vitro studies of plasma sprayed hydroxyapatite coatings. Powder Diffraction 26:138-143

336. Oguchi H, Ishikawa K, Ojima S, Hirayama Y, Seto K, Eguchi G (1992) Evaluation of a high-velocity flame-spraying technique for hydroxyapatite. Biomaterials 13:471-477

337. Ohring M (2002) Materials science of thin films. Academic, San Diego, CA, USA. p 794

338. Okido M, Kuroda K, Ishikawa M, Ichino R, Takai O (2002) Hydroxyapatite coating on titanium by means of thermal substrate method in aqueous solutions. Solid State Ion 151:47-52

339. Oliveira AL, Elvira C, Reis RL, Vazquez B, San Roman J (1999) Surface modification tailors the characteristics of biomimetic nucleated on starch-based polymers. J Mater Sci Mater Med 10:827-835

340. Oliveira AL, Mano JF, Reis RL (2003) Nature-inspired calcium phosphate coatings: present status and novel advances in the science of mimicry. Curr. Opin. Solid State Mater. Sci. 7:309-318

341. Ong JL, Chan DCN (1999) Hydroxyapatite and their use as coatings in dental implants: a review. Crit Rev Biomed Eng 28:667-707

342. Ong JL, Lucas LC (1994) Post-deposition heat treatment for ion beam sputter deposited calcium phosphate coatings. Biomaterials 15:337-341

343. Ong JL, Harris LA, Lucas LC, Lacefield WR, Rigney D (1991) X-ray photoelectron spectroscopy characterization of ion beam sputter deposited calcium phosphate coatings. J Am Ceram Soc 74:2301-2304

344. Ong JL, Harris LA, Lucas LC, Lacefield WR, Rigney D (1992) Structure, solubility and bond strength of thin calcium phosphate coatings produced by ion beam sputter-deposited. Biomaterials 13:249-254

345. Ong JL, Lucas LC, Raikar GN, Weimer JJ, Gregory JC (1994) Surface characterization of ion beam sputter-deposited Ca-P coatings after in vitro immersion. Coll Surf A 87:151-162

346. Ong JL, Bessho K, Cavin R, Carnes DL (2002) Bone response to radio frequency sputtered calcium phosphate implants and titanium implants in vivo. J Biomed Mater Res 59:184-190

347. Onoki T, Hashida T (2006) New method for hydroxyapatite coating of titanium by the hydrothermal hot isostatic pressing technique. Surf CoatTechnol 200:6801-6807

348. Oosterbos CJM, Vogely HC, Nijhof MW, Fleer A, Verbout AJ, Tonino AJ, Dhert WJA (2002) Osseointegration of hydroxyapatite-coated and noncoated Ti6Al4V implants in the presence of local infection: a comparative histomorphometrical study in rabbits. J Biomed Mater Res 60:339-347

349. Oosterbos CJM, Rahmy AIA, Tonino AJ, Witpeerd W (2004) High survival rate of hydroxyapatite-coated hip prostheses 100 consecutive hips followed for 10 years. Acta Orthop. Scandinavica 75:127-133

350. Ozeki K, Yuhta T, Aoki H, Nishimura I, Fukui Y (2001) Push-out strength of hydroxyapatite coated by sputtering technique in bone. Bio-Med. Mater. Eng. 11:63-68

351. Ozeki K, Yuhta T, Aoki H, Fukui Y (2003) Inhibition of Ni release from NiTi alloy by hydroxyapatite, alumina, and titanium sputtered coatings. Bio-Med. Mater. Eng. 13:271-279

352. Ozeki K, Fukui Y, Aoki H (2007) Influence of the calcium phosphate content of the target on the phase composition and deposition rate of sputtered films. Appl Surf Sci 253:5040-5044

353. Ozeki K, Aoki H, Masuzawa T (2010) Influence of the hydrothermal temperature and pH on the crystallinity of a sputtered hydroxyapatite film. Appl Surf Sci 256:7027-7031

354. Paital SR, Dahotre NB (2007) Laser surface treatment for porous and textured Ca-P bio-ceramic coating on Ti-6Al-4 V. Biomed Mater 2:274-281

355. Paital SR, Dahotre NB (2008) Review of laser based biomimetic and bioactive Ca-P coatings. Mater Sci Technol 24:1144-1161

356. Paital SR, Dahotre NB (2009) Calcium phosphate coatings for bio-implant applications: materials, performance factors, and methodologies. Mater Sci Eng R 66:1-70

357. Paital SR, Dahotre NB (2009) Wettability and kinetics of hydroxyapatite precipitation on a laser-textured Ca–P bioceramic coating. Acta Biomater 5:2763-2772

358. Paital SR, Balani K, Agarwal A, Dahotre NB (2009) Fabrication and evaluation of a pulse laser-induced Ca-P coating on a Ti alloy for bioapplication. Biomed Mater 4:015009

359. Park S, Condrate R, Hoelzer DT, Fischman GS (1998) Interfacial characterization of plasma-spray coated calcium phosphate on Ti-6Al-4 V. J Mater Sci Mater Med 9:643-649

360. Park YS, Yi KY, Lee IS, Han CH, Jung YC (2005) The effects of ion beam-assisted deposition of hydroxyapatite on the grit-blasted surface of endosseous implants in rabbit tibiae. Int J Oral Maxillofac Implants 20:31-38

361. Pawlowski L (2008) The science and engineering of thermal spray coatings. Wiley, New York, USA. p 691

362. Peng P, Kumar S, Voelcker NH, Szili E, Smart RSC, Griesser HJ (2006) Thin calcium phosphate coatings on titanium by electrochemical deposition in modified simulated body fluid. J Biomed Mater Res A 76A:347-355

363. Peraire C, Arias JL, Bernal D, Pou J, León B, Arañó A, Roth W (2006) Biological stability and osteoconductivity in rabbit tibia of pulsed laser deposited hydroxylapatite coatings. J Biomed Mater Res A 77A:370-379

364. Pezeshki P, Lugowski S, Davies JE (2010) Dissolution behavior of calcium phosphate nanocrystals deposited on titanium alloy surfaces. J Biomed Mater Res A 94A:660-666

365. Pham MT, Matz W, Grambole D, Herrmann F, Reuther H, Richter E, Steiner G (2002) Solution deposition of hydroxyapatite on titanium pre-treated with a sodium ion implantation. J Biomed Mater Res 59:716-724

366. Piattelli A, Cosci F, Scarano A, Trisi P (1995) Localized chronic suppurative bone infection as a sequel of peri-implantitis in a hydroxyapatite-coated dental implant. Biomaterials 16:917-920

367. Pilliar RM, Deporter DA, Watson PA, Pharoah M, Chipman M, Valiquette N, Carter S, de Groot K (1991) The effect of partial coating with hydroxyapatite on bone remodeling in relation to porous-coated titanium-alloy dental implants in the dog. J Dent Res 70:1338-1345

368. Podlesak H, Pawlowski L, D'Haese R, Laureyns J, Lampke T, Bellayer S (2010) Advanced microstructural study of suspension plasma sprayed hydroxyapatite coatings. J Thermal Spray Technol 19:657-664

369. Pontin MG, Lange FF, Sanchez-Herencia AJ, Moreno R (2005) Effect of unfired tape porosity on surface film formation by dip coating. J Am Ceram Soc 88:2945-2948

370. Prymak O, Bogdansky D, Esenwein SA, Köller M, Epple M (2004) NiTi shape memory alloys coated with calcium phosphate by plasma-spraying. Chemical and biological properties. Mater Werkst 35:346-351

371. Quaranta A, Iezzi G, Scarano A, Coelho PG, Vozza I, Marincola M, Piattelli A (2010) A histomorphometric study of nanothickness and plasma-sprayed calcium-phosphorous-coated implant surfaces in rabbit bone. J. Periodontology 81:556-561

372. Quek CH, Khor KA, Cheang P (1999) Influence of processing parameters in the plasma spraying of hydroxyapatite/Ti-6Al-4V composite coatings. J Mater Process Technol 89–90:550-555

373. Rabiei A, Thomas B, Jin C, Narayan R, Cuomo J, Yang Y, Ong JL (2006) A study on functionally graded HA coatings processed using ion beam assisted deposition with in situ heat treatment. Surf CoatTechnol 200:6111-6116

374. Rajaratnam SS, Jack C, Tavakkolizadeh A, George MD, Fletcher RJ, Hankins M, Shepperd JAN (2008) Long-term results of a hydroxyapatite-coated femoral component in total hip replacement: a 15- to 21-year follow-up study. J. Bone Joint Surg. B 90:27-30

375. Rajesh P, Muraleedharan CV, Komath M, Varma H (2011) Pulsed laser deposition of hydroxyapatite on titanium substrate with titania interlayer. J Mater Sci Mater Med 22:497-505

376. Rau JV, Smirnov VV, Laureti S, Generosi A, Varvaro G, Fosca M, Ferro D, Cesaro SN, Albertini VR, Barinov SM (2010) Properties of pulsed laser deposited fluorinated hydroxyapatite films on titanium. Mater Res Bull 45:1304-1310

377. Rautray TR, Narayanan R, Kwon TY, Kim KH (2010) Surface modification of titanium and titanium alloys by ion implantation. J Biomed Mater Res B Appl Biomater 93B:581-591

378. Redepenning J, Schlessinger T, Burnham S, Lippiello L, Miyano J (1996) Characterization of electrolytically prepared brushite and hydroxyapatite coatings on orthopedic alloys. J Biomed Mater Res 30:287-294

379. Reis RL, Monteiro FJ, Hastings GW (1994) Stability of hydroxylapatite plasma-sprayed coated Ti–6Al–4V under cyclic bending in simulated physiological solution. J Mater Sci Mater Med 5:457-462

380. Roguska A, Hiromoto S, Yamamoto A, Woźniak MJ, Pisarek M, Lewandowska M (2011) Collagen immobilization on 316 L stainless steel surface with cathodic deposition of calcium phosphate. Appl Surf Sci 257:5037-5045

381. Rossler S, Sewing A, Stolzel M, Born R, Scharnweber D, Dard M, Worch H (2003) Electrochemically assisted deposition of thin calcium phosphate coatings at near-physiological pH and temperature. J Biomed Mater Res A 64A:655-663

382. Roy M, Bandyopadhyay A, Bose S (2011) Induction plasma sprayed nano hydroxyapatite coatings on titanium for orthopaedic and dental implants. Surf CoatTechnol 205:2785-2792

383. Ruckenstein E, Gourisankar SV (1986) Preparation and characterization of thin film surface coatings for biological environments. Biomaterials 7:403-422

384. Saithna A (2010) The influence of hydroxyapatite coating of external fixator pins on pin loosening and pin track infection: a systematic review. Injury 41:128-132

385. Saju KK, Reshmi R, Jayadas NH, James J, Jayaraj MK (2009) Polycrystalline coating of hydroxyapatite on TiAl6V4 implant material grown at lower substrate temperatures by hydrothermal annealing after pulsed laser deposition. Proc Inst Mech Eng H 223:1049-1057

386. Schliephake H, Scharnweber D, Roesseler S, Dard M, Sewing A, Aref A (2006) Biomimetic calcium phosphate composite coating of dental implants. Int J Oral Max Impl 21:738-746

387. Schwarz MLR, Kowarsch M, Rose S, Becker K, Lenz T, Jani L (2009) Effect of surface roughness, porosity, and a resorbable calcium phosphate coating on osseointegration of titanium in a minipig model. J Biomed Mater Res A 89A:667-678

388. Shetty AA, Slack R, Tindall A, James KD, Rand C (2005) 1 Results of a hydroxyapatite-coated (Furlong) total hip replacement. A 13- to 15-year follow-up. J. Bone Joint Surg. B 87:1050-1054

389. Shi JZ, Chen CZ, Yu HJ, Zhang SJ (2008) Application of magnetron sputtering for producing bioactive ceramic coatings on implant materials. Bull Mater Sci 31:877-884

390. Shirkhanzadeh M (1993) electrochemical preparation of bioactive calcium phosphate coatings on porous substrates by the periodic pulse technique. J Mater Sci Lett 12:16-19

391. Shirkhanzadeh M (1998) direct formation of nanophase hydroxyapatite on cathodically polarized electrodes. J Mater Sci Mater Med 9:67-72

392. Siebers MC, Walboomers XF, Leeuwenburgh SCG, Wolke JGC, Jansen JA (2004) Electrostatic spray deposition (ESD) of calcium phosphate coatings, an in vitro study with osteoblast-like cells. Biomaterials 25:2019-2027

393. Siebers MC, Matsuzaka K, Walboomers XF, Leeuwenburgh SCG, Wolke JGC, Jansen JA (2005) Osteoclastic resorption of calcium

phosphate coatings applied with electrostatic spray deposition (ESD), in vitro. J Biomed Mater Res A 74A:570-580

394. Siebers MC, Walboomers XF, Leewenburgh SCG, Wolke JCG, Boerman OC, Jansen JA (2006) Transforming growth factor-β1 release from a porous electrostatic spray deposition-derived calcium phosphate coating. Tiss. Eng. 12:2449-2456

395. Siebers MC, Wolke JGC, Walboomers FX, Leeuwenburgh SCG, Jansen JA (2007) In vivo evaluation of the trabecular bone behavior to porous electrostatic spray deposition-derived calcium phosphate coatings. Clin. Oral Implant. Res. 18:354-361

396. Silva MHPD, Lima JHC, Soares GA, Elias CN, de Andrade MC, Best SM, Gibson IR (2001) Transformation of monetite to hydroxyapatite in bioactive coatings on titanium. Surf Coat Technol 137:270-276

397. Sima LE, Stan GE, Morosanu CO, Melinescu A, Ianculescu A, Melinte R, Neamtu J, Petrescu SM (2010) Differentiation of mesenchymal stem cells onto highly adherent radio frequency-sputtered carbonated hydroxylapatite thin films. J Biomed Mater Res A 95A:1203-1214

398. Simank HG, Stuber M, Frahm R, Helbig L, van Lenthe H, Müller R (2006) The influence of surface coatings of dicalcium phosphate (DCPD) and growth and differentiation factor-5 (GDF-5) on the stability of titanium implants in vivo. Biomaterials 27:3988-3994

399. Singh RK, Narayan J (1990) Pulsed-laser evaporation technique for deposition of thin films: physics and theoretical model. Phys. Rev. B 41:8843-8859

400. Singh RK, Qian F, Nagabushnam V, Damodaran R, Moudgil BM (1994) Excimer laser deposition of hydroxylapatite thin films. Biomaterials 15:522-528

401. Snyders R, Bousser E, Music D, Jensen J, Hocquet S, Schneider JM (2008) Influence of the chemical composition on the phase constitution and the elastic properties of RF-sputtered hydroxyapatite coatings. Plasma Processes Polym 5:168-174

402. Søballe K, Hansen ES, Brockstedt-Rasmussen HB, Hjortdal VE, Juhl GI, Pedersen CM, Hvid I, Bünger C (1991) Gap healing

enhanced by hydroxyapatite coatings in dogs. Clin Orthop 272:300-307

403. Søballe K, Hansen ES, Brockstedt-Rasmussen HB, Bünger C (1993) Hydroxyapatite coating converts fibrous tissue to bone around loaded implants. J. Bone Jt. Surg. 75B:270-278

404. Sobolev VV, Guilemany JM (1996) Dynamic processes during high velocity oxyfuel spraying. Int Mater Rev 41:13-32

405. Socol G, Torricelli P, Bracci B, Iliescu M, Miroiu F, Bigi A, Werckmann J, Mihailescu IN (2004) Biocompatible nanocrystalline octacalcium phosphate thin films obtained by pulsed laser deposition. Biomaterials 25:2539-2545

406. Song WH, Jun YK, Han Y, Hong SH (2004) Biomimetic apatite coatings on micro-arc oxidized Titania. Biomaterials 25:3341-3349

407. Sousa SR, Barbosa MA (1996) Effect of hydroxyapatite thickness on metal ion release from Ti6Al4V substrates. Biomaterials 17:397-404

408. Sridhar TM, Kamachi MU, Subbaiyan M (2003) Sintering atmosphere and temperature effects on hydroxyapatite coated type 316 L stainless steel. Corr Sci 45:2337-2359

409. Stephenson PK, Freeman MAR, Revell PA, Germain J, Tuke M, Pirie CJ (1991) The effect of hydroxyapatite coating on growth of bone into cavities in an implant. J Arthroplasty 6:51-58

410. Stevenson JR, Solnick-Legg H, Legg KO (1989) Production of high adherent hydroxyapatite coatings by ion beam and plasma techniques. Mater Res Soc Symp Proc 110:715-719

411. Stilling M, Rahbek O, Søballe K (2009) Inferior survival of hydroxyapatite versus titanium-coated cups at 15 years. Clin. Orthop. Rel. Res. 467:2872-2879

412. Stoch A, Brozek A, Kmita G, Stoch J, Jastrzebski W, Rakowska A (2001) Electrophoretic coating of hydroxyapatite on titanium implants. J Mol Struct 596:191-200

413. Stoica TF, Morosanu C, Slav A, Stoica T, Osiceanu P, Anastasescu C, Gartner M, Zaharescu M (2008) Hydroxyapatite films obtained by sol–gel and sputtering. Thin Solid Films 516:8112-8116

414. Suchanek WL, Yoshimura M (1998) Processing and properties of hydroxyapatite-based biomaterials for use as hard tissue replacement implants. J Mater Res 13:94-117

415. Sudo SZ, Schotzko NK, Folke LEA (1976) Use of hydroxyapatite coated glass beads for preclinical testing of potential antiplaque agents. Appl Environmental Microbiology 32:428-437

416. Sun T, Wang M (2010) Electrochemical deposition of apatite/collagen composite coating on NiTi shape memory alloy and coating properties. Mater Res Soc Symp Proc 1239:141-146

417. Sun L, Berndt CC, Gross KA, Kucuk A (2001) Material fundamentals and clinical performance of plasma sprayed hydroxyapatite coatings: a review. J Biomed Mater Res (Appl Biomater) 58:570-592

418. Sun L, Berndt CC, Grey CP (2003) Phase, structural and microstructural investigations of plasma sprayed hydroxyapatite coatings. Mater Sci Eng, A 360:70-84

419. Sun J, Han Y, Huang X (2007) Hydroxyapatite coatings prepared by micro-arc oxidation in Ca- and P-containing electrolyte. Surf Coat Technol 201:5655-5658

420. Surmenev RAA (2012) Review of plasma-assisted methods for calcium phosphate-based coatings fabrication. Surf Coat Technol 206:2035-2056

421. Surmenev RA, Ryabtseva MA, Shesterikov EV, Pichugin VF, Peitsch T, Epple M (2010) The release of nickel from nickel-titanium (NiTi) is strongly reduced by a sub-micrometer thin layer of calcium phosphate deposited by rf-magnetron sputtering. J Mater Sci Mater Med 21:1233-1239

422. Suzuki M, Calasans-Maia MD, Marin C, Granato R, Gil JN, Granjeiro JM, Coelho PG (2010) Effect of surface modifications on early bone healing around plateau root form implants: an experimental study in rabbits. J Oral Maxillofac Surg 68:1631-1638

423. Sygnatowicz M, Tiwari A (2009) Controlled synthesis of hydroxyapatite-based coatings for biomedical application. Mater Sci Eng C 29:1071-1076

424. Takadama H, Kim HM, Kokubo T, Nakamura T (2001) TEM-EDX study of mechanism of bonelike apatite formation on bioactive

titanium metal in simulated body fluid. J Biomed Mater Res 57:441-448

425. Tas AC, Bhaduri SB (2004) Rapid coating of Ti6Al4V at room temperature with a calcium phosphate solution similar to 10x simulated body fluid. J Mater Res 19:2742-2749

426. Thomas KA, Cook CD, Ray RJ, Jarcho M (1989) Biologic response to hydroxylapatite coated titanium hips. J Arthroplasty 4:43-53

427. Tieanboon P, Jaruwangsanti N, Kiartmanakul S (2009) Efficacy of hydroxyapatite in pedicular screw fixation in canine spinal vertebra. Asian Biomedicine 3:177-181

428. Tinsley D, Watson CJ, Russell JL (2001) A comparison of hydroxylapatite coated implant retained fixed and removable mandibular prostheses over 4 to 6 years. Clin Oral Implant Res 12:159-166

429. Tkalcec E, Sauer M, Nonninger R, Schmidt H (2001) Sol–gel-derived hydroxyapatite powders and coatings. J Mater Sci 36:5253-5263

430. Tong W, Chen J, Zhang X (1995) Amorphorization and recrystallization during plasma spraying of hydroxyapatite. Biomaterials 16:829-832

431. Tong W, Yang Z, Zhang X, Yang A, Feng J, Cao Y, Chen J (1998) Studies on diffusion maximum in X-ray diffraction patterns of plasma-sprayed hydroxyapatite coatings. J Biomed Mater Res 40:407-413

432. Toque JA, Hamdi M, Ide-Ektessabi A, Sopyan I (2009) Effect of the processing parameters on the integrity of calcium phosphate coatings produced by RF-magnetron sputtering. Int. J. Modern Phys. B 23:5811-5818

433. Torrisi L, Setola R (1993) thermally assisted hydroxyapatite obtained by pulsed-laser deposition on titanium substrates. Thin Solid Films 227:32-36

434. Tri LQ, Chua DHC (2009) an investigation into the effects of high laser fluence on hydroxyapatite/calcium phosphate films deposited by pulsed laser deposition. Appl Surf Sci 256:76-80

435. Trisi P, Keith DJ, Rocco S (2005) Human histologic and histomorphometric analyses of hydroxyapatite-coated implants after 10 years of function: a case report. Int J Oral Maxillofac Implants 20:124-130

436. Tsui YC, Doyle C, Clyne TW (1998) Plasma sprayed hydroxyapatite coatings on titanium substrates. Part 1: Mechanical properties and residual stress levels. Biomaterials 19:2015-2029

437. Tucker BE, Cottel CM, Auyeung RCY, Spector M, Nancollas GH (1996) Pre-conditioning and dual constant composition dissolution kinetics of pulsed laser deposited hydroxyapatite thin films on silicon substrates. Biomaterials 17:631-637

438. Uchida M, Kim HM, Kokubo T, Fujibayashi S, Nakamura T (2003) Structural dependence of apatite formation on Titania gels in a simulated body fluid. J Biomed Mater Res A 64A:164-170

439. Ueda K, Narushima T, Goto T, Taira M, Katsube T (2007) Fabrication of calcium phosphate films for coating on titanium substrates heated up to 773 K by RF magnetron sputtering and their evaluations. Biomed Mater 2:S160-S166

440. Ueda K, Kawasaki Y, Narushima T, Goto T, Kurihara J, Nakagawa H, Kawamura H, Taira M (2009) Calcium phosphate films with/ without heat treatments fabricated using RF magnetron sputtering. J Biomech Sci Eng 4:392-403

441. U.S. Food and Drug Administration (1995)http://www.fda.gov/MedicalDevices/DeviceRegulationandGuidance/GuidanceDocuments/ucm080224.htm webcite.

442. Valter S, Orfeo S, Clarke DR (1997) Mechanical and chemical consequences of the residual stresses in plasma sprayed hydroxyapatite coatings. Biomaterials 18:477-482

443. van der Wal E, Wolke JGC, Jansen JA, Vredenberg AM (2005) Initial reactivity of rf magnetron sputtered calcium phosphate thin films in simulated body fluids. Appl Surf Sci 246:183-192

444. Van der Wal E, Oldenburg SJ, Heij T, van der Gon AWD, Brongersma HH, Wolke JGC, Jansen JA, and Vredenberg AM (2006) Adsorption and desorption of Ca and PO4 species from SBFs on RF-sputtered calcium phosphate thin films. Appl Surf Sci 252:3843-3854

445. Van Dijk K, Schaeken HG, Wolke JGC, Maree CHM, Habraken FHPM, Verhoven J, Jansen JA (1995) Influence of discharge power level on the properties of hydroxyapatite films deposited on Ti6Al4V with RF magnetron sputtering. J Biomed Mater Res 29:269-276

446. Van Dijk K, Schaeken HG, Wolke JGC, Jansen JA (1996) Influence of annealing temperature on RF magnetron sputtered calcium phosphate coatings. Biomaterials 17:405-410

447. van Dijk K, Verhoeven J, Marée CHM, Habraken FHPM, Jansen JA (1997) Study of the influence of oxygen on the composition of thin films obtained by r.f. sputtering from a Ca5(PO4)3OH target. Thin Solid Films 304:191-195

448. Variola F, Brunski JB, Orsini G, de Oliveira TP, Wazen R, Nanci A (2011) Nanoscale surface modifications of medically relevant metals: State-of-the art and perspectives. Nanoscale 3:335-353

449. Vasanthan A, Kim H, Drukteinis S, Lacefield W (2008) Implant surface modification using laser guided coatings: in vitro comparison of mechanical properties. J. Prosthodontics 17:357-364

450. Vercaigne S, Wolke JGC, Naert I, Jansen JA (2000) A mechanical evaluation of TiO2-gritblasted and Ca-P magnetron sputter coated implants placed into the trabecular bone of the goat: Part 1. Clin. Oral Implant. Res. 11:305-313

451. Vercaigne S, Wolke JGC, Naert I, Jansen JA (2000) A histological evaluation of TiO2-gritblasted and Ca-P magnetron sputter coated implants placed into the trabecular bone of the goat: Part 2. Clin. Oral Implant. Res. 11:314-324

452. Verestiuc L, Morosanu C, Bercu M, Pasuk I, Mihailescu IN (2004) Chemical growth of calcium phosphate layers on magnetron sputtered HA films. J. Cryst. Growth 264:483-491

453. Walschus U, Hoene A, Neumann HG, Wilhelm L, Lucke S, Luthen F, Rychly J, Schlosser M (2009) Morphometric immunohistochemical examination of the inflammatory tissue reaction after implantation of calcium phosphate-coated titanium plates in rats. Acta Biomater 5:776-784

454. Wan T, Aoki H, Hikawa J, Lee JH (2007) RF-magnetron sputtering technique for producing hydroxyapatite coating film on various substrates. Bio-Med. Mater. Eng. 17:291-297

455. Wang L, Nancollas GH (2008) Calcium orthophosphates: crystallization and dissolution. Chem Rev 108:4628-4669

456. Wang G, Zreiqat H (2010) Functional coatings or films for hard-tissue applications. Materials 3:3994-4050

457. Wang BC, Lee TM, Chang E, Yang CY (1993) The shear strength and the failure mode of plasma-sprayed hydroxyapatite coating to bone: the effect of coating thickness. J Biomed Mater Res 27:1315-1327

458. Wang CK, Lin JHC, Ju CP, Ong HC, Chang RPH (1997) Structural characterization of pulsed laser-deposited hydroxyapatite film on titanium substrate. Biomaterials 18:1331-1338

459. Wang XX, Hayakawa S, Tsuru K, Osaka A (2000) A comparative study of in vitro apatite deposition on heat-, H2O2-, and NaOH-treated titanium surfaces. J Biomed Mater Res 52:172-178

460. Wang CX, Chen ZQ, Guan LM, Wang M, Liu ZY, Wang PL (2001) Fabrication and characterization of graded calcium phosphate coatings produced by ion beam sputtering/mixing deposition. Nucl Instr Methods Phys Res B 179:364-372

461. Wang C, Ma J, Cheng W, Zhang R (2002) Thick hydroxyapatite coatings by electrophoretic deposition. Mater Lett 57:99-105

462. Wang XX, Yan W, Hayakawa S, Tsuru K, and Osaka a (2003) Apatite deposition on thermally and anodically oxidized titanium surfaces in a simulated body fluid. Biomaterials 24:4631-4637

463. Wang J, Layrolle P, Stigter M, de Groot K (2004) Biomimetic and electrolytic calcium phosphate coatings on titanium alloy: physicochemical characteristics and cell attachment. Biomaterials 25:583-592

464. Wang XJ, Li YC, Lin JG, Hodgson PD, Wen CE (2008) Apatite-inducing ability of titanium oxide layer on titanium surface: the effect of surface energy. J Mater Res 23:1682-1688

465. Wang DG, Chen CZ, Ma J, Zhang G (2008) In situ synthesis of HA coating by laser cladding. Coll. Surf. B 66:155-162

466. Wang X, Li Y, Lin J, Hodgson PD, Wen C (2008) Effect of heat-treatment atmosphere on the bond strength of apatite layer on Ti substrate. Dental Mater. 24:1549-1555

467. Wang C, Gross KA, Anderson GI, Dunstan CR, Carbone A, Berger G, Ploska U, Zreiqat H (2009) Bone growth is enhanced by novel bioceramic coatings on Ti alloy implants. J Biomed Mater Res A 90A:419-428

468. Wang LJ, Lu JW, Xu FS, Zhang FS (2011) Dynamics of crystallization and dissolution of calcium orthophosphates at the near-molecular level. Chinese Sci. Bull. 56:713-721

469. Wei M, Ruys AJ, Milthorpe BK, Sorrell CC, Evans JH (2001) Electrophoretic deposition of hydroxyapatite coatings on metal substrates: a nanoparticulate dual-coating approach. J Sol–gel Part Sci Technol 21:39-48

470. Wei D, Zhou Y, Wang Y, Jia D (2007) Characteristic of microarc oxidized coatings on titanium alloy formed in electrolytes containing chelate complex and nano-HA. Appl Surf Sci 253:5045-5050

471. Wen HB, Wolke JGC, de Wijn JR, Liu Q, Cui FZ (1997) Fast precipitation of calcium phosphate layers on titanium induced by simple chemical treatments. Biomaterials 18:1471-1478

472. Weng W, Baptisa JL (1998) Alkoxide route for preparing hydroxyapatite and its coatings. Biomaterials 19:125-131

473. Weng J, Liu X, Zhang X, Ma Z, Ji X, Zyman Z (1993) Further studies on the plasma-sprayed amorphous phase in hydroxyapatite coatings and its deamorphization. Biomaterials 14:578-582

474. Wheeler SL (1996) Eight-year clinical retrospective study of titanium plasma-sprayed and hydroxyapatite-coated cylinder implants. Int J Oral Maxillofac Implants 11:340-350

475. Wheeler DL, Montfort MJ, McLoughlin SW (2001) Differential healing response of bone adjacent to porous implants coated with hydroxyapatite and 45 S5 bioactive glass. J Biomed Mater Res 55:603-612

476. Wie H, Hero H, Solheim T (1998) Hot isostatic pressing-processed hydroxyapatite-coated titanium implants: light microscopic and scanning electron microscopy investigations. Int J Oral Maxillofac Implants 13:837-844

477. Wikipedia (2012a) Coating.http://en.wikipedia.org/wiki/Coating e May 2012.

478. Wikipedia (2012b) Electrophoretic deposition.http://en.wikipedia.org/wiki/Electrophoretic_deposition May 2012.

479. Willmann G (1999) Coating of implants with hydroxyapatite – material connections between bone and metal. Adv Eng Mater 1:95-105

480. Willmott PR, Huber JR (2000) Pulsed laser vaporization and deposition. Rev. Modern Phys. 72:315-328

481. Wolke JGC, de Blieck-Hogervorst JMA, Dhert WJA, Klein CPAT, de Groot K (1992) Studies on thermal spraying of apatite bioceramics. J Thermal Spray Technol 1:75-82

482. Wolke JGC, van Dijk K, Schaeken HG, de Groot K, Jansen JA (1994) Study of the surface characteristics of magnetron-sputter calcium phosphate coatings. J Biomed Mater Res 28:1477-1484

483. Wolke JGC, van der Waerden JPCM, de Groot K, Jansen JA (1997) Stability of radiofrequency magnetron sputtered calcium phosphate coatings under cyclically loaded conditions. Biomaterials 18:483-488

484. Wolke JGC, de Groot K, Jansen JA (1998) In vivo dissolution behavior of various RF magnetron sputtered Ca-P coatings. J Biomed Mater Res 39:524-530

485. Wolke JGC, van der Waerden JPCM, Schaeken HG, Jansen JA (2003) In vivo dissolution behavior of various RF magnetron-sputtered Ca-P coatings on roughened titanium implants. Biomaterials 24:2623-2629

486. Wu GM, Hsiao WD, Kung SF (2009) Investigation of hydroxyapatite coated polyether ether ketone composites by gas plasma sprays. Surf Coat Technol 203:2755-2758

487. Xu S, Long J, Sim L, Diong CH, Ostrikov K (2005) RF plasma sputtering deposition of hydroxyapatite bioceramics: synthesis, performance, and biocompatibility. Plasma Processes Polym 2:373-390

488. Xue W, Liu X, Zheng X, Ding C (2005) Effect of hydroxyapatite coating crystallinity on dissolution and osseointegration in vivo. J Biomed Mater Res A 74A:553-561

489. Yamaguchi T, Tanaka Y, Ide-Ektessabi A (2006) Fabrication of hydroxyapatite thin films for biomedical applications using RF magnetron sputtering. Nucl. Instrum. Methods Phys. Res. B 249:723-725

490. Yamashita K, Arashi T, Kitagaki K, Yamada S, Umegaki T (1994) Preparation of apatite thin films through RF-sputtering from calcium phosphate glasses. J Am Ceram Soc 77:2401-2407

491. Yan Y, Wolke JGC, de Ruijter A, Li Y, Jansen JA (2006) Growth behavior of rat bone marrow cells on RF magnetron sputtered hydroxyapatite and dicalcium pyrophosphate coatings. J Biomed Mater Res A 78A:42-49

492. Yang YC (2011) Investigation of residual stress generation in plasma-sprayed hydroxyapatite coatings with various spraying programs. Surf CoatTechnol 205:5165-5171

493. Yang Y, Chang E (2005) Measurements of residual stresses in plasma-sprayed hydroxyapatite coatings on titanium alloy. Surf CoatTechnol 190:122-131

494. Yang CY, Wang BC, Chang E, Wu JD (1995) The influences of plasma spraying parameters on the characteristics of hydroxyapatite coatings: a quantitative study. J Mater Sci Mater Med 6:249-257

495. Yang CY, Wang BC, Chang WJ, Chang E, Wu JD (1996) Mechanical and histological evaluations of cobalt-chromium alloy and hydroxyapatite plasma-sprayed coatings in bone. J Mater Sci Mater Med 7:167-174

496. Yang CY, Wang BC, Lee TM, Chang E, Chang GL (1997) Intramedullary implant of plasma-sprayed hydroxyapatite coating: an interface study. J Biomed Mater Res 36:39-48

497. Yang YC, Chang E, Hwang BH, Lee SY (2000) biaxial residual stress states of plasma-sprayed hydroxyapatite coatings on titanium alloy substrate. Biomaterials 21:1327-1337

498. Yang Y, Agarwal CM, Kim KH, Martin H, Schul K, Bumgardner JM, Ong JL (2003) Characterization and dissolution behavior of sputtered calcium phosphate coatings after different postdeposition heat treatment temperatures. J Oral Implantology 29:270-277

499. Yang Y, Kim KH, Agarwal CM, Ong JL (2003) Effect of post-deposition heating temperature and the presence of water vapor during heat treatment on crystallinity of calcium phosphate coatings. Biomaterials 24:5131-5137

500. Yang YC, Chang E, Lee SY (2003) Mechanical properties and Young's modulus of plasma-sprayed hydroxyapatite coating on Ti substrate in simulated body fluid. J Biomed Mater Res A 67A:886-899

501. Yang Y, Kim KH, Ong JL (2005) A review on calcium phosphate coatings produced using a sputtering process – an alternative to plasma spraying. Biomaterials 26:327-337

502. Yang JX, Jiao YP, Cui FZ, Lee IS, Yin QS, Zhang Y (2008) Modification of degradation behavior of magnesium alloy by IBAD coating of calcium phosphate. Surf CoatTechnol 202:5733-5736

503. Yang CW, Lui TS, Chen LH (2009) hydrothermal crystallization effect on the improvement of erosion resistance and reliability of plasma-sprayed hydroxyapatite coatings. Thin Solid Films 517:5380-5385

504. Yang S, Man HC, Xing W, Zheng X (2009) Adhesion strength of plasma-sprayed hydroxyapatite coatings on laser gas-nitrided pure titanium. Surf CoatTechnol 203:3116-3122

505. Yang S, Xing W, Man HC (2009) Pulsed laser deposition of hydroxyapatite film on laser gas nitriding NiTi substrate. Appl Surf Sci 255:9889-9892

506. Yen SK, Lin CM (2002) cathodic reactions of electrolytic hydroxyapatite coating on pure titanium. Mater Chem Phys 77:70-76

507. Yoon HJ, Song JE, Um YJ, Chae GJ, Chung SM, Lee IS, Jung UW, Kim CS, Choi SH (2009) Effects of calcium phosphate coating to SLA surface implants by the ion-beam-assisted deposition method on self-contained coronal defect healing in dogs. Biomed Mater 4:044107

508. Yoshinari M, Ohshiro Y, Derand T (1994) thin hydroxyapatite coating produced by the ion beam dynamic mixing method. Biomaterials 15:529-535

509. Yoshinari M, Watanabe Y, Ohtsuka Y, Dérand T (1997) Solubility control of thin calcium-phosphate coating with rapid heating. J Dent Res 76:1485-1494

510. Yoshinari M, Oda Y, Inoue T, Matsuzaka K, Shimono M (2002) Bone response to calcium phosphate-coated and bisphosphonate-immobilized titanium implants. Biomaterials 23:2879-2885

511. You C, Yeo IS, Kim MD, Eom TK, Lee JY, Kim S (2005) Characterization and in vivo evaluation of calcium phosphate coated cp-titanium by dip-spin method. Curr Appl Phys 5:501-506

512. Yousefpour M, Afshar A, Yang X, Li X, Yang B, Wu Y, Chen J, Zhang X (2006) Nano-crystalline growth of electrochemically deposited apatite coating on pure titanium. J Electroanal Chem 589:96-105

513. Yuan Q, Sahu LK, D'Souza NA, Golden TD (2009) Synthesis of hydroxyapatite coatings on metal substrates using a spincasting technique. Mater Chem Phys 116:523-526

514. Zalm PC (1989) Quantitative sputtering. In: Cuomo JJ, Rossnagel SM, Kaufman HR (eds) Handbook of ion beam processing technology, Noyes Publications; Park Ridge, NJ, USA. pp 78-111

515. Zeng H, Lacefield WR (2000) The study of surface transformation of pulsed laser deposited hydroxyapatite coatings. J Biomed Mater Res 50:239-247

516. Zeng H, Lacefield WR, Mirov S (2000) Structural and morphological study of pulsed laser deposited calcium phosphate bioceramic coatings: influence of deposition conditions, laser parameters, and target properties. J Biomed Mater Res 50:248-258

517. Zhang L, Zhang WT (2011) Numerical investigation on particle velocity in cold spraying of hydroxyapatite coating. Adv Mater Res 188:717-722

518. Zhang S, Xianting Z, Yongsheng W, Kui C, Wenjian W (2006) Adhesion strength of sol–gel derived fluoridated hydroxyapatite coatings. Surf Coat Technol 200:6350-6354

519. Zhang L, Zhang W, Wu Z (2012) Numerical simulation of hydroxyapatite particle impacting on Ti substrate in cold spraying. Appl. Mech. Mater. 130–134:900-903

520. Zhang L, Zhang W, Li H, Geng W, Bao Y (2012) Development of a cold spraying system for fabricating hydroxyapatite coating. Appl. Mech. Mater. 151:300-304

521. Zhao Z, Li H, Huo G, Sun P, Li Y, Kuroda K, Ichino R, Okido M (2003) Cathodic deposition of hydroxyapatite without H2 evolution. TMS Annual Meeting. pp 169-173

522. Zhao BH, Lee IS, Bai W, Cui FZ, Feng HL (2005) Improvement of fibroblast adherence to titanium surface by calcium phosphate coating formed with IBAD. Surf Coat Technol 193:366-371

523. Zhao GL, Wen G, Song Y, Wu K (2011) Near surface martensitic transformation and recrystallization in a Ti-24Nb-4Zr-7.9Sn alloy substrate after application of a HA coating by plasma spraying. Mater Sci Eng C 31:106-113

524. Zheng M, Fan D, Li XK, Zhang JB, Liu QB (2010) Microstructure and in vitro bioactivity of laser-cladded bioceramic coating on

titanium alloy in a simulated body fluid. J. Alloys Compounds 489:211-214

525. Zhitomirsky I (2000) Electrophoretic hydroxyapatite coatings and fibers. Mater Lett 42:262-271

526. Zhitomirsky I, Gal-Or L (1997) Electrophoretic deposition of hydroxyapatite. J Mater Sci Mater Med 8:213-219

527. Zhu Y, Chen Y, Xu G, Ye X, He D, Zhong J (2012) Micropattern of nano-hydroxyapatite/silk fibroin composite onto Ti alloy surface via template-assisted electrostatic spray deposition. Mater Sci Eng C 32:390-394

528. Ziani-Cherif H, Abe Y, Imachi K, Matsuda T (2002) Hydroxyapatite coating on titanium by thermal substrate method in aqueous solution. J Biomed Mater Res 59:390-397

529. Zyman Z, Weng J, Liu X, Zhang X, Ma Z (1993) Amorphous phase and morphological structure of hydroxyapatite plasma coatings. Biomaterials 14:225-228

530. Zyman Z, Weng J, Liu X, Li X, Zhang X (1994) Phase and structural changes in hydroxyapatite coatings under heat treatment. Biomaterials 15:151-155

Multicolor Layer-by-Layer Films Using Weak Polyelectrolyte Assisted Synthesis of Silver Nanoparticles

Pedro Jose Rivero, Javier Goicoechea, Aitor Urrutia,
Ignacio Raul Matias, and Francisco Javier Arregui

Nanostructured Optical Devices Laboratory, Electric and Electronic
Engineering Department, Public University of Navarra, Edif.Los Tejos,
Campus Arrosadía, 31006, Pamplona, Spain

ABSTRACT

In the present study, we show that silver nanoparticles (AgNPs) with different shape, aggregation state and color (violet, green, orange) have been successfully incorporated into polyelectrolyte multilayer thin films using the layer-by-layer (LbL) assembly. In order to obtain colored thin films based on AgNPs is necessary to maintain the aggregation state of the nanoparticles, a non-trivial aspect in which this work is focused on. The use of Poly(acrylic acid, sodium salt) (PAA) as a protective agent of the AgNPs is the key element to preserve the aggregation state and makes possible the presence of similar aggregates (shape and size) within the LbLcolored films. This approach based on electrostatic interactions of the polymeric chains and the immobilization of AgNPs with different shape and size into the thin films opens up a new interesting perspective to fabricate multicolornanocomposites based on AgNPs.

BACKGROUND

The synthesis of metal nanoparticles (gold, silver, palladium, copper) and their further incorporation into thin films is of great interest for applications in antibacterial coatings [1,2], catalysis [3,4], chemical sensors [5,6], drug delivery [7,8], electronics [9], photochemistry [10] or photonics [11,12]. The wide variety of synthesis methodologies to obtain the metallic particles provide alternative ways to synthesize the nanoparticles controlling various parameters such as the shape, size, surface functionalization or interparticle distance which affect their final properties. A control of these parameters is a challenging goal, and a large number of reports have been published [13-20]. Among them, the synthesis routes based on the chemical reduction in organic solvents or in which polymers can act simultaneously as a stabilizer and reducer agent to obtain metal nanoparticles have been investigated [21,23]. However, the use of organic media and the synthesis of polydisperse nanoparticles limit their use for some specific applications in where monodisperse nanoparticles are required [24,25].

Alternative procedures for the synthesis of Au or AgNPs are based on the use of water soluble polymers with the aim of achieving size-controlled nanoparticles. Wang and co-workers have obtained AuNPs in aqueous solution in the 1–5 nm size range with the use of poly(methacrylic acid) (PMMA) [26,27]. Keuker-Baumann and co-workers reported a study about the formation of AgNPs with a high control and a characteristic plasmon band at 410 nm is observed using dilute solutions of long-chain sodium polyacrylates (NaPA) by exposing the solutions to UV-radiation [28] in where the coil size of the polymeric chains acts as a collector of silver cations (Ag+). Other researches have investigated the formation of AgNPs and intermediate clusters in polyacrylate aqueous solutions by chemical reduction of Ag + using a reducing agent, gamma radiation or ambient light[29-32]. Very recently, our group has described the synthesis of multicolor silver nanoparticles with a high stability in time, using poly(acrylic acid, sodium salt) (PAA) as a protective agent, in where the AgNPs exhibit localized surface plasmon resonance (LSPR) spectra (colors) as a function of variable protective and reducing agents with a well-defined shape and size [33].

Once the metallic nanoparticles have been synthesized, a further assembly in the form of thin films is required to obtain the desired silver nanoparticle composites. However, this is not always possible because of the need of preserving the aggregation state of the nanoparticles. Several approaches are based on the incorporation of the nanoparticles into a previous polymeric matrix obtained by different thin film techniques, such as sol–gel deposition or electrospinning process[34,35]. In all the cases, the presence of an intense absorption band at 410 nm is indicative of spherical AgNPs with a characteristic yellow coloration. In this work, layer-by-layer (LbL) assembly allows to manipulate and incorporate the nanoparticles into the thin films due to the use of PAA as a protective agent which maintains unaltered the aggregation state of the AgNPs. This technique is based on the alternating deposition of oppositely charged polyelectrolytes in water solution (polycations and polyanions) on substrates where the electrostatic interaction between these two components of different charge is the driving force for the multilayer assembly [36]. Previous works are based on the in situ synthesis of AgNPs in the polyelectrolyte multilayers via counterion exchange and posterior reduction [37-41]. In these cases, this approach is based on the pH-dependent dissociation of weak acids such as PAA as a function of the pH, in where both ionized (carboxylate) and non-ionized (carboxylic) groups are obtained. The presence of the free ionic groups makes possible to bind metal ions via a simple aqueous ion exchange procedure and a posterior chemical reduction step with a reducing agent, leads to obtain the nanoparticles within the thin film. However, Su and co-workers have demonstrated the incorporation of AgNPs with the use of strong polyelectrolytes, such as poly(diallyldimethylammonium chloride) (PDDA) and poly(styrene sulfonate) (PSS), without any further adjustment of the pH [42]. Although the film thickness of the polymeric matrix can be perfectly controlled by the number of layers deposited onto the substrate, a better control over particles size and distribution in the films are not easy to achieve with the in situ chemical reduction and as a result, only yellow coloration is observed. Our hypothesis for obtaining the color is due to a greater degree control over particles (shape and size distribution) in the films with a real need of maintaining the aggregation state.

To overcome this situation, we propose a first stage of synthesis of multicolorAgNPs (violet, green and orange) in aqueous polymeric solution (PAA) with a well-defined shape and size. A second stage is based on the incorporation of these AgNPs into a polyelectrolyte multilayer thin film using the layer-by-layer (LbL) assembly. To our knowledge, this is the first time that a study about the color formation based on AgNPs is investigated in films preserving the original color of the solutions.

METHODS

Materials

Poly(allylamine hydrochloride) (PAH) (Mw 56,000), Poly(acrylic acid, sodium salt) 35 wt% solution in water (PAA) (Mw 15,000), silver nitrate (>99% titration) and boranedimethylamine complex (DMAB) were purchased from Sigma-Aldrich and used without any further purification.

SYNTHESIS METHOD OF THE PAA-CAPPED AGNPS

Multicolor silver nanoparticles have been prepared by adding freshly variable DMAB concentration (0.033, 0.33 and 3.33 mM) to vigorously stirred solution which contained constant PAA (25 mM) and $AgNO_3$ concentrations (3.33 mM). This yields a molar ratio between the protective and loading agent ([PAA]/[AgNO$_3$] ratio of 7.5:1. The final molar ratios between the reducing and loading agents ([DMAB]/[AgNO$_3$] ratio) were 1:100, 1:10 and 1:1. The reduction of silver cations (Ag$^+$) and all subsequent experiments were performed at room conditions and stored at room temperature. More details of this procedure can be found in the literature [33].

Fabrication of the Multilayer Film

Aqueous solutions of PAH and PAA with a concentration of 25 mM with respect to the repetitive unit were prepared using ultrapure deionized water (18.2 MΩ·cm). The pH was adjusted to 7.5 by the addition of a few drops of NaOH or HCl. The LbL assembly was performed by sequentially exposing the glass slide (substrate) to cationic polyelectrolyte poly(allylamine hydrochloride) (PAH) and anionic polyelectrolyte PAA loaded with the silver nanoparticles previously synthesized (PAA-Ag NPs) with an immersion time of 5 minutes. A rinsing step of 1 minute in deionized water was performed between the two polyelectrolytes baths and a drying step of 30 seconds was performed after each rinsing step. The combination of a cationic monolayer with an anionic monolayer is called bilayer. The LbL process was carried out using a 3-axis cartesian robot from Nadetech Innovations. More details of the LbL assembly can be found elsewhere [35,36,43]. No atmospheric oxidation of the LbL films with AgNPs was observed using this experimental process, showing the long-term stability of the resultant films.

Characterization

UV-visible spectroscopy (UV–vis) was used to characterize the optical properties of the multicolor silver nanoparticles and the resultant coatings obtained by LbL assembly. Measurements were carried out with a Jasco V-630 spectrophotometer.

Transmission electron microscopy (TEM) was used to determine the morphology (shape and size) of the silver nanoparticles obtained in aqueous solution. This TEM analysis was carried out with a Carl Zeiss Libra 120. Samples for TEM were prepared by dropping and evaporating the solutions onto a collodion-coated copper grid.

Atomic force microscope (AFM) in tapping mode (Innova, Veeco Inc.) has been used in order to show the distribution of the Ag NPs, thickness and roughness of the films obtained by the LbL assembly.

RESULTS AND DISCUSSION

In Figure 1, it is possible to appreciate three different colors obtained (violet, green and orange) using PAA as an encapsulating agent (PAA-AgNPs) when DMAB concentration is increased (from 0,033 mM to 3.33 mM). These poly(acrylic acid)-coated nanoparticles are unique in this respect because prior studies using different encapsulating agents to synthesize silver nanoparticles indicate that only an orange coloration is obtained without any color variation. In addition, the resultant PAA-AgNPs dispersions showed an excellent long-term stability since no changes in the position of their absorption bands have been observed after more than one year of storage at room conditions, corroborated by UV–vis spectroscopy.

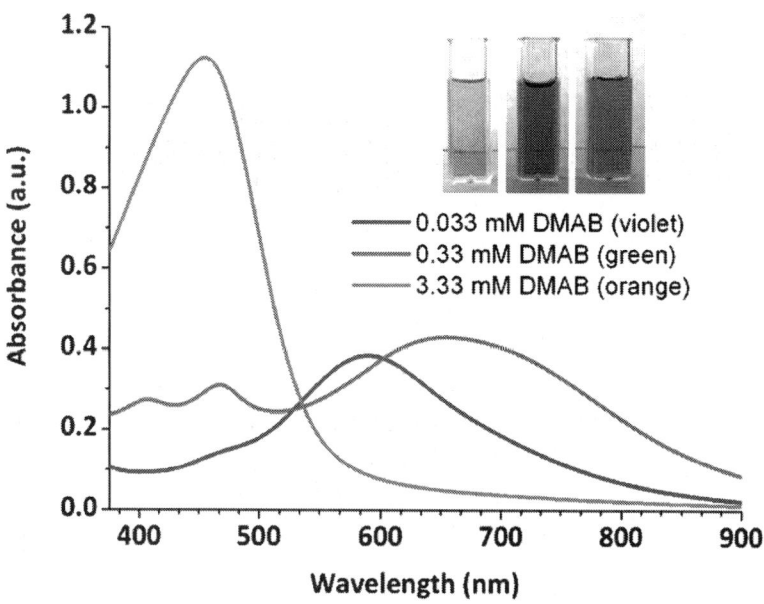

Figure 1: UV–vis spectroscopy of the multicolor silver nanoparticles (violet, green, orange) as a function of DMAB concentration.

Initially, the mixture of 25 mM PAA with $AgNO_3$ is colorless (control), but after the addition of 0.033 mM of DMAB, the mixture turns quickly to violet with a plasmonic absorption peak with a maximum centered at 600 nm. When DMAB concentration is increased (0.33 mM), the sample changes from violet to green. The absorption band distribution in the UV–VIS spectrum was altered significantly. The initial absorption band was increased significantly, and it was also shifted toward longer-wavelengths (at 650 nm). Furthermore, a new absorption band was found at 480 nm related with the coexistence of different Ag-NP aggregation states or shapes. Finally, when DMAB concentration is increased to 3.33 mM, the solution turned to orange color and only an intense absorption band around 440 nm was observed, indicating the complete synthesis of spherical silver nanoparticles. The evolution of these absorption bands in two well separated regions (region 1 for the 400–500 nm and region 2 for the 600–700 nm) has been discussed in previous works [33]. These changes in the UV–vis spectra (colors) are related to changes in the shape, size and aggregation state of the AgNPs. In order to corroborate this hypothesis, TEM analysis of the different samples (PAA-AgNPs) were performed (see Figure 2).

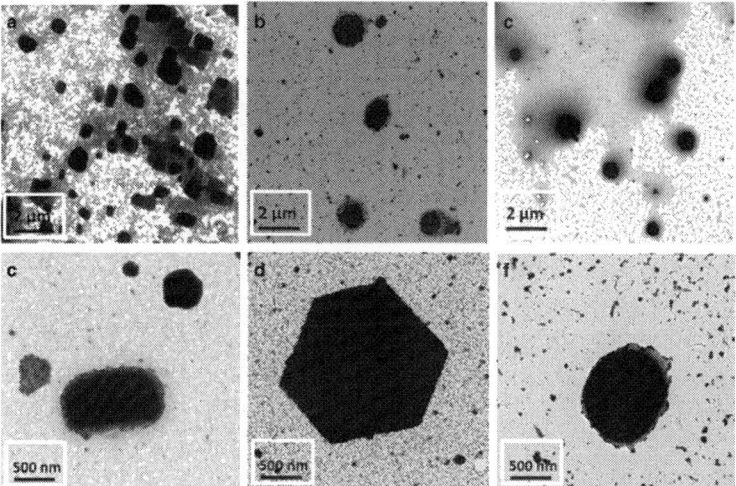

Figure 2: TEM micrographs of the multicolor silver nanoparticles at different scale (500 nm and 2 μm). (a,d) rod shape (violet coloration); (b,e) hexagonal shape (green coloration); (c,f) spherical shape (orange coloration).

According to the results observed in Figures 1 and 2, when DMAB concentration added in the reaction mixture is low, violet coloration ($[DMAB]/[AgNO_3] = 0.01$) or green coloration ($[DMAB]/[AgNO_3] = 0.1$) is observed with a typical long-wavelength absorption band (600–700 nm) and a new absorption band at 480 nm appears for green coloration, which corresponds to complexes of small positively charged metal clusters and polymer ligands of the polyacrylate anions (PAA) [44-46]. It has been also found that AgNPs with a specific shape and size (TEM micrographs), nanorods of different size (from 100 to 500 nm) are synthesized for violet coloration. Additionally, clusters with a hexagonal shape (from 0.5-1 μm) mixed with spherical particles of nanometricsize are found for green coloration. However, when DMAB concentration is increased ($[DMAB]/[AgNO3] = 1$), orange coloration with an intense absorption band at 440 nm is observed, which is indicative of a total reduction of the silver cations and the corresponding synthesis of spherical nanoparticles with variable size. These results corroborate that the excess of free Ag^+cations immobilized into the polyelectrolyte chains of the PAA respect to the reducing agent, plays a key role in the synthesis process, yielding different nanoparticle size distributions and aggregation states. It is important to remark that changes in the plasmonic absorption bands (resultant color) basically depend on the relationship between the aggregation state of the nanoparticles (even in the cluster formation) and the final shape/size of the resultant nanoparticles. A control of all these parameters is the key to understand the color formation in the films.

The next step is to incorporate the previously synthesized colored AgNPs in a polyelectrolyte multilayer film using the layer-by-layer (LbL) assembly. The main goal is to get a coating with the similar coloration that the initial colored solution of PAA-AgNPs (violet, green and orange). Therefore, it is necessary to maintain the aggregation state of the nanoparticles into the thin film. Then, the multilayer assembly of both cationic polyelectrolyte poly(allylamine hydrochloride) (PAH) and anionic polyelectrolyte PAA loaded with the AgNPs previously synthesized (colored PAA-Ag NPs) depends on the degree of ionization of the polymers and their charge density which is perfectly controlled with a suitable adjustment of the pH [47,48]. An important consideration of this work is that the deposition of PAH and PAA-AgNPs is at the same pH (7.5) because PAA at this pH or higher pH values plays a key role in order to preserve the aggregation state of the nanoparticles during the synthesis process (Figure 3) with a perfect control of the resultant color without any further precipitation. When the pH of the dipping solutions (PAA-AgNPs) is lowered below 7.0, a change of the coloration is observed in all the experiments which it is indicative of a loss of the aggregation state of the PAA-AgNPs with an increase in opalescence and a further precipitation with a complete loss of color (transparent solutions) at low pH values (pH 4.0 or lower).

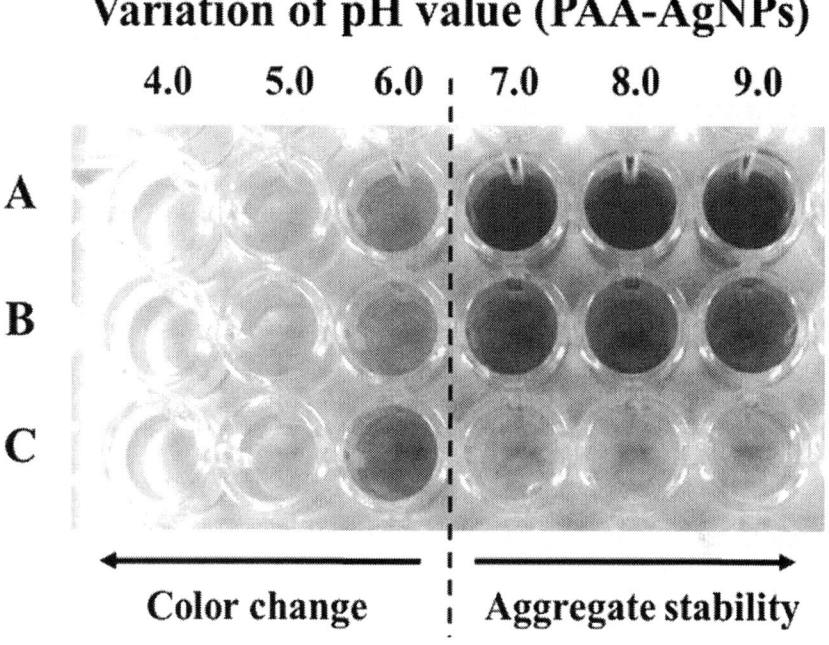

Figure 3: Variation of the multicolor silver nanoparticles (PAA-AgNPs) as a function of the pH value for violet (A), green (B) and orange coloration (C).

Due to these changes concerning to the color as a function of the pH dipping solutions, the reason of choosing pH 7.5 for both PAH and PAA-AgNPs is the base to obtain the multicolor films. In addition, the fundamental element to obtain the multilayer buildup is the presence of ionized groups of these weak polyelectrolytes, which are responsible for the electrostatic assembly and the spatial control of the previously silver nanoparticles distribution (colored PAA-AgNPs) in the multilayer film when the number of bilayers is increased. In Figure 4, a detail of the evolution of the absorption peaks (UV–vis spectroscopy) and the corresponding color formation during the LbL fabrication process for both PAH and PAA-AgNPs (orange coloration) is shown as a function of the number of bilayers added to the corresponding films.

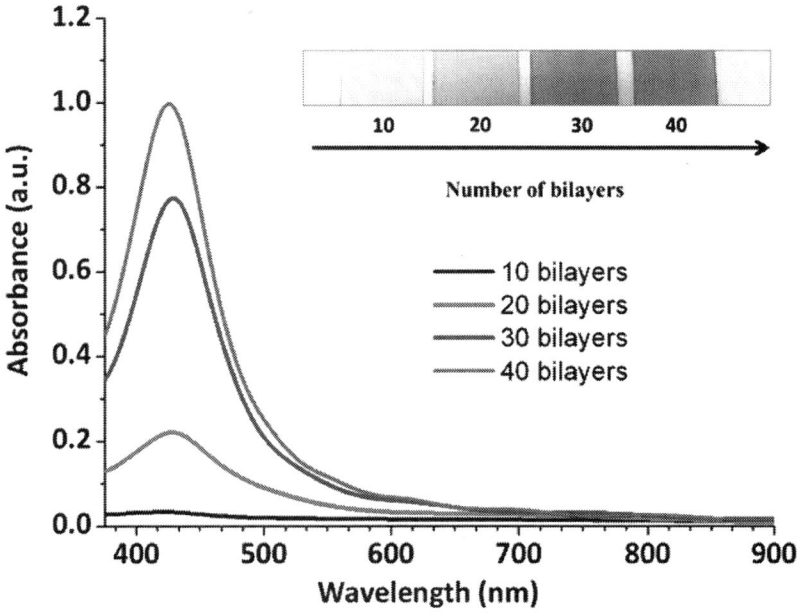

Figure 4: UV–vis spectroscopy of the orange multilayer films for different number of bilayers (10, 20, 30 and 40) and photographs of the coatings.

From the results of Figure 4, it can be said that a successful deposition of orange colored films was obtained. A LSPR absorption peak centred at 440 nm grows as a function of the number of bilayers deposited onto glass slides via LbL assembly (10, 20, 30 and 40 bilayers, respectively). The intensity increase of the absorption band at 440 nm or the orange coloration of the films , is the result of an incorporation of spherical AgNPs in the multilayer assembly.

As it has been previously commented, the aim of this manuscript is to get thin films with the same coloration that the initial PAA-AgNPs solution. The next step will be to incorporate the violet silver nanoparticles in the LbLbuildup. In Figure 5, a study of the evolution of the absorption bands corresponding to both PAH and PAA-AgNPs (violet) during the LbL fabrication process is studied at the same number of bilayers.

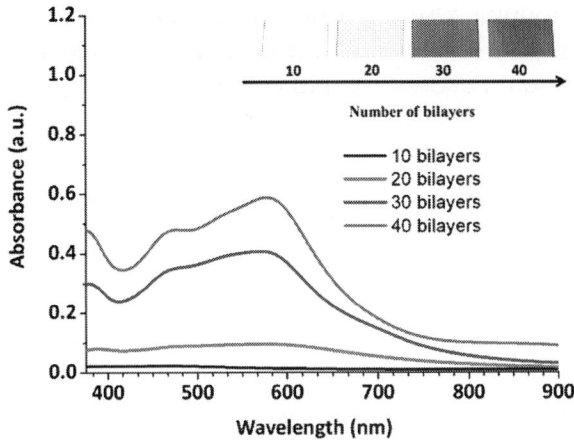

Figure 5: UV–vis spectroscopy of the violet multilayer films for different number of bilayers (10, 20, 30 and 40) and photographs of the coatings.

According to the results, an increase of the absorption peak from 10 bilayers to 40 bilayers at a specific wavelength position is observed. The location of this absorption band, which is higher in intensity when the thickness of the coating is increased, maintains the same position that initial synthesized violet silver nanoparticles (PAA-AgNPs) at 600 nm (see Figure 1). In view of these results, UV–vis spectra reveal identical absorption peaks for both LbL fabrication process and the synthesized PAA-AgNPs (violet solution), which it means that silver nanoparticles with a specific shape (mostly rods) have been successfully incorporated in the multilayer assembly.

In Figure 6, the evolution of the absorption bands corresponding to the coating of PAH and PAA-AgNPs (green) during LbL fabrication process is shown. UV–vis spectra of the resulting coatings at different number of bilayers confirm the existence of two absorption peaks during the multilayer assembly, one at 640 nm typical of green AgNPs which is lower in intensity and the other one, higher in intensity at 440 nm. For this case, it is possible to appreciate a difference in the UV–vis spectra between the LbL multilayer assembly and the previously green colored PAA-AgNPs (see Figure 1). In the opinion of the authors, the presence of a higher and broader absorption band at 440 nm is due to an agglomeration and higher number of the AgNPs inside of the thin film and the presence of AgNPs with different shape (not only hexagonal shape). This approach is corroborated by the final coloration of the resultant coatings in where a light orange coloration instead of clearly green coloration is observed. A possible reason of this spectral change (color) in comparison with previously PAA-AgNPs could be associated to the reduction of the metal clusters with a partial positive charge by the amine groups [49,50] of the PAH during the LbL assembly. However, this hypothesis has not been observed for the violet coloration (Figure 5) when the number of bilayers onto glass slides was continuously increased, so we can conclude that a reduction by the amine groups of PAH and a further in situ generation of the spherical AgNPs is not observed. According to the results, the presence of the absorption band at 440 nm is associated to the incorporation of AgNPs with less size (mostly spherical nanoparticles) during the fabrication process (observed by TEM images), whereas the absorption band at 480 nm is lower in intensity because of a more difficult incorporation of higher size particles (metal clusters with hexagonal shape) in the multilayer films for a total number of 40 bilayers. As conclusion, we can remark that a selective absorption process is observed and as result, it is the partial orange coloration of the resultant films due to a higher presence of spherical AgNPs in comparison with hexagonal clusters.

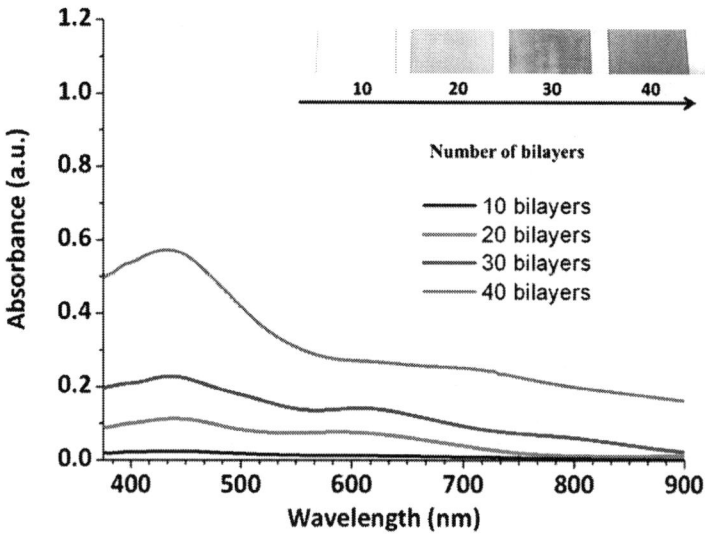

Figure 6: UV–vis spectroscopy of the green multilayer films for different number of bilayers (10, 20, 30 and 40) and photographs of the coatings.

In order to understand the incorporation of the multicolorAgNPs inside the LbL assembly, the position of the absorption bands with their corresponding intensities and the aspect in coloration of the final films have been analyzed. However, to create a template of well-defined coloration, the thickness of the resulting films to incorporate the AgNPs plays a key role, which is perfectly controlled by two factors, the pH value of the polyelectrolyte solutions (PAH and PAA-AgNPs) and the number of bilayers deposited onto glass slides [47,48]. When the pH of the dipping solutions is 7.5, both PAH and PAA-AgNPs are adsorbed as fully charged polyelectrolytes and very thin films are obtained. For a total of 40 bilayers, the average thickness is varied from 185 nm (PAH/PAA-AgNPs violet coating), 223 nm (PAH/PAA-AgNPs orange coating) to 293 nm (PAH/PAA-AgNPs green coating). In Figure 7, the evolution of the thickness for different number of bilayers (10, 20, 30 and 40, respectively) with their error bars in this pH regime (7.5) is shown. According to these thickness results, it is possible to appreciate that PAH/PAA-AgNPs with a light orange coloration instead of clearly green coloration is due to the higher incorporation of AgNPs with nanometric spherical size instead of metal clusters into the film for a coating of 40 bilayers.

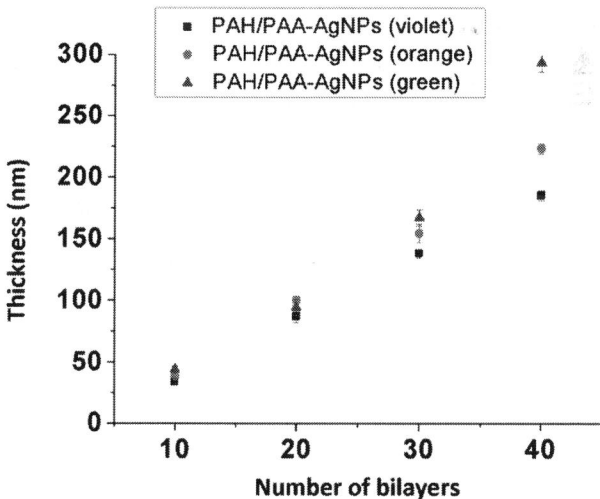

Figure 7: Evolution of thickness of the PAH/PAA-AgNPs multilayer assemblies (violet, green, orange) for different number of bilayers.

Obviously, in all the cases of study, the thickness and the resultant color formation depends basically on surface charge of both ionized PAH/PAA polymeric chains, the number of bilayers deposited, the number of the AgNPs incorporated and the distribution of them with a specific shape during the fabrication process. In order to show the aspect of the thin films after LbL fabrication process, AFM images of 40 bilayers [PAH/PAA-AgNPs] at pH 7.5 reveal that the morphologies of the thin films were homogeneous, very slight porous surfaces with an average roughness (rms) of 12.9 nm (violet coloration), 16.7 nm (green coloration) and 18.6 nm (orange coloration). In all the cases, the polymeric chains of the weak polyelectrolytes (PAH and PAA) are predominant in the outer surface and the AgNPs are embedded inside the polymeric films. In order to show the presence of these AgNPs in the LbL assembly, a thermal treatment of the films was necessary with the idea of evaporating the polymeric chains (PAH and PAA, respectively) and so, the contribution of the AgNPs can be appreciated when the fabrication process is performed.

In Figure 8, AFM images corresponding to 10, 20, 30 and 40 bilayers of PAH/PAA-AgNPs (violet coloration) after a thermal treatment of 450°C are shown. In all the images, the only presence of AgNPs is observed and a complete change in the morphology is observed for all the films when the number of bilayers was increased. Initially, when the coating has 10 bilayers it is possible to appreciate well-separated AgNPs with a very low roughness of 5.8 nm. However, when the number of bilayers is increased, the roughness is changing from 10.2 nm (20 bilayers) to 23.9 nm (30 bilayers) and 28.7 nm (40 bilayers). It is important to remark that after a thermal treatment, the total evaporation of the polymeric chains induces an agglomeration of the AgNPs without preserving their distribution along the films. This aspect is corroborated due to a color change from violet to orange in the resultant films.

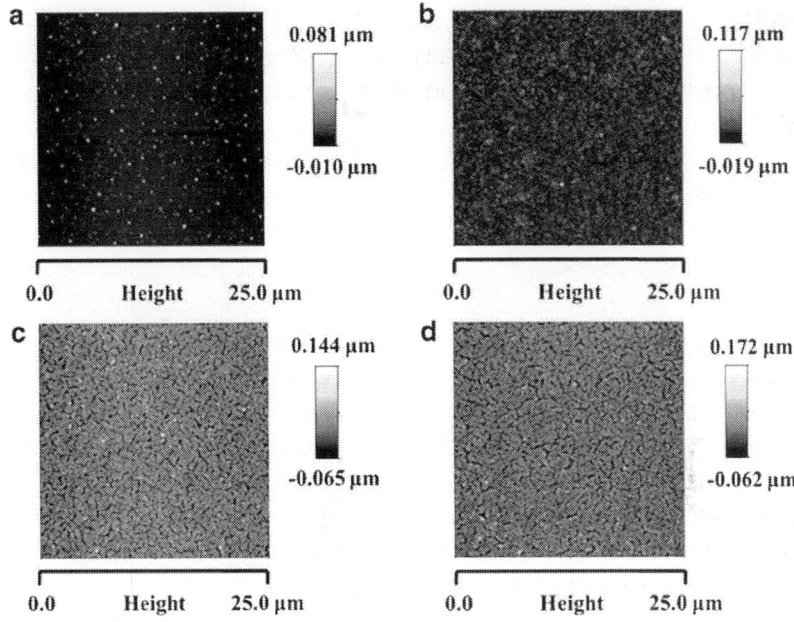

Figure 8: AFM images (25x25 **μ**m) of PAH/PAA-AgNPs (violet coloration) after a thermal treatment as a function of number of bilayers (a) 10 bilayers; (b) 20 bilayers; (c) 30 bilayers and (d) 40 bilayers.

In other words, the fact that a higher number of bilayers during the LbL fabrication process, and consequently, a higher thickness of the resultant films, promote a better definition of the color, mostly in the green coloration (see Figure 9) because of a better entrapment of both initial clusters (hexagons with higher size) and nanometric spherical AgNPs in the multilayer assembly. Additionally, new PAH/PAA-AgNPs coatings of 80 bilayers at pH 7.5 have been fabricated in order to show clearly the final coloration onto the glass slides as a function of the initial synthesized multicolor silver nanoparticles (PAA-AgNPs).

Figure 9: Final aspect of the PAH/PAA-AgNPs multilayer assembly (violet, green, orange coloration) for a total number of 80 bilayers.

Figure 10 shows the UV–vis spectra of the samples prepared with this thickness (80 bilayers) and the spectra reveal that the position of the absorption bands is the same than previous spectra (Figures 3, 4 and 5) but with a considerable increase in intensity of the absorption peaks due to a higher number of the metallic silver nanoparticles that have been incorporated into the multilayer film. Therefore, when the thickness is increased, it is possible to corroborate the presence of the same aggregates species or AgNPs than the original colloidal solutions. In other words, when the thickness is increased, the final coloration of the resultant films (violet, green or orange) is similar than the color of the original colloidal PAA-AgNPs solutions. These results of coloration as a function of number bilayers indicate that a higher thickness leads to a better incorporation of higher size aggregates (clusters) in the resultant films. This is the first time that a study about colored AgNPs synthesis and their incorporation in multicolor films (violet, green or orange) is investigated using the LbL assembly. These multicolor LbL films can be used for optical fiber sensor applications [41]. The retention of the color of the Ag-colloidal dispersion in the LbL films makes possible the fabrication of optical fiber sensors with optical responses related to their specific LSPR absorption bands. In such case, different optical fiber sensor signals could be multiplexed into a single optical fiber enabling multipoint measurement.

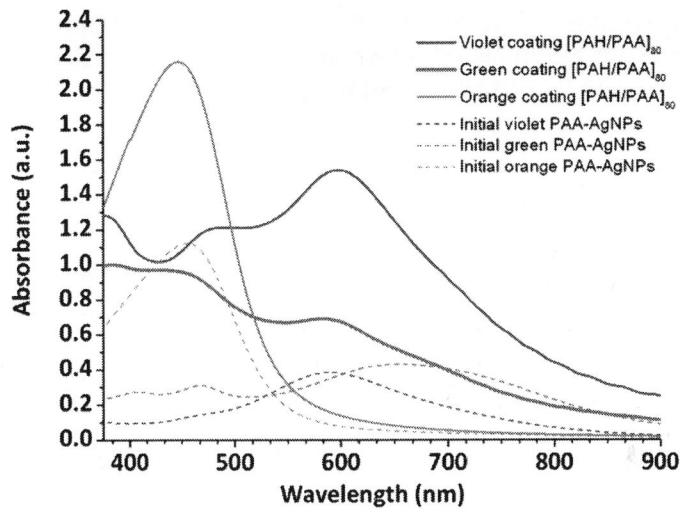

Figure 10: UV–vis spectra of the multilayer thin films of 80 bilayers PAH/PAA-AgNPs (violet, green and orange coloration) in comparison with initial colored PAA-AgNPs solutions.

CONCLUSIONS

In this work, highly stable coloredAgNPs were synthesized using a water-based synthesis route using PAA as capping agent. The weak polyelectrolyte nature of the PAA and the excess of Ag+cations respect to the concentration of reducing agent (DMAB) make possible to achieve nanoparticles with different sizes, shapes and aggregation states. This yields different coloredAgNPs dispersions (violet, green and orange). Such AgNPs have been successfully incorporated into LbL thin films in where the adsorption process was carried out that the AgNPs and aggregates (clusters) within the film are maintained, and thus the coloration of the films is also kept. In order to obtain the proper coloration of the thin film, a study about the influence of the number of PAH/PAA-AgNPs bilayers added (10, 20, 30, 40 and 80, respectively), the position of the absorption bands (UV–vis spectra) and the pH value of the weak polyelectrolytes solutions have been performed. A pH value of 7.5 or higher value of the PAA-AgNPs solution is the key to preserve the aggregation state of the AgNPs without any further precipitation or loss of coloration. A better definition of the coloration in the films is observed when a higher number of bilayers (thickness) are added during the LbL assembly (mostly in green color) because of a better entrapment of both initial clusters and nanometric spherical nanoparticles. This is indicative of a higher number of AgNPs or aggregates of specific shape and size that are incorporated into the multilayer film. In addition, AFM images reveal a low roughness of the resultant colored films which drastically changes with a thermal treatment due a total evaporation of the polymeric chains (PAH and PAA), making possible to appreciate the number of AgNPs incorporated as a function of bilayers added. To our knowledge, this is the first time that colored PAA-AgNPs of different sizes and shapes are synthesized and incorporated later in LbL assemblies preserving the original color of the solutions.

AUTHORS' CONTRIBUTIONS

PJR carried out the main part of the experimental work, and carried out the syntehsis process of the coatings. He participated in the design of the study and in the draft of the manuscript. JG participated in the experimental work, carried out the AFM measurements and contributed with the draft of the manuscript. AU participated in the experimental work and carried out the UV–vis spectra. IRM participated in the design of the study and helped to draft the manuscript. FJA participated in the design of the study and helped to draft the manuscript. All authors read and approved the final manuscript.

ACKNOWLEDGMENTS

This work was supported in part by the Spanish Ministry of Economy and Competitiveness CICYT FEDER TEC2010-17805 research grant. The authors express their gratitude to David García-Ros (Universidad de Navarra) for his help with the TEM images.

REFERENCES

1.   Abdullayev E, Sakakibara K, Okamoto K, Wei W, Ariga K, Lvov Y: Natural tubule clay template synthesis of silver nanorods for antibacterial composite coating. *ACS Appl Mater Interfaces* 2011, 3:4040-4046

2.   Malcher M, Volodkin D, Heurtault B, André P, Schaaf P, Möhwald H, Voegel J, Sokolowski A, Ball V, Boulmedais F, Frisch B: Embedded silver ions-containing liposomes in polyelectrolyte multilayers: cargos films for antibacterial agents. *Langmuir* 2008, 24:10209-10215

3.   Kidambi S, Bruening ML: Multilayered polyelectrolyte films containing palladium nanoparticles: synthesis, characterization, and application in selective hydrogenation. *Chem Mater* 2005, 17:301-307

4.   Kidambi S, Dai J, Li J, Bruening ML: Selective Hydrogenation by Pd nanoparticles embedded in polyelectrolyte multilayers. *J Am Chem Soc* 2004, 126:2658-2659.

5.   Xi Q, Chen X, Evans DG, Yang W: Gold nanoparticle-embedded porous graphene thin films fabricated via layer-by-layer self-assembly and subsequent thermal annealing for electrochemical sensing. *Langmuir* 2012, 28:9885-9892.

6.   Devadoss A, Spehar-Délèze A, Tanner DA, Bertoncello P, Marthi R, Keyes TE, Forster RJ:Enhanced electrochemiluminescence and charge transport through films of metallopolymer-gold nanoparticle composites. *Langmuir* 2010, 26:2130-2135.

7.   Dreaden EC, Alkilany AM, Huang X, Murphy CJ, El-Sayed MA: The golden age: Gold nanoparticles for biomedicine. *Chem Soc Rev* 2012, 41:2740-2779.

8.   Doane TL, Burda C: The unique role of nanoparticles in nanomedicine: Imaging, drug delivery and therapy. *Chem Soc Rev* 2012, 41:2885-2911.

9.   Shang L, Wang Y, Huang L, Dong S: Preparation of DNA-silver nanohybrids in multilayer nanoreactors by in situ electrochemical reduction, characterization, and application. *Langmuir* 2007, 23:7738-7744.

10. Logar M, Jančar B, Šturm S, Suvorov D: Weak polyion multilayer-assisted in situ synthesis as a route toward a plasmonic Ag/TiO2 photocatalyst. *Langmuir* 2010, 26:12215-12224.

11. Nolte AJ, Rubner MF, Cohen RE: Creating effective refractive index gradients within polyelectrolyte multilayer films: molecularly assembled rugate filters. *Langmuir* 2004, 20:3304-3310.

12. Wang TC, Cohen RE, Rubner MF: Metallodielectric photonic structures based on polyelectrolyte multilayers. *Adv Mater* 2002, 14:1534-1537. Publisher Full Text

13. Vigderman L, Khanal BP, Zubarev ER: Functional gold nanorods: synthesis, self-assembly, and sensing applications. *Adv Mater* 2012, 24:4811-4841.

14. Jeon S, Xu P, Zhang B, MacK NH, Tsai H, Chiang LY, Wang H: Polymer-assisted preparation of metal nanoparticles with controlled size and morphology. *J Mat Chem* 2011, 21:2550-2554.

15. Cobley CM, Skrabalak SE, Campbell DJ, Xia Y: Shape-controlled synthesis of silver nanoparticles for plasmonic and sensing applications. *Plasmonics* 2009, 4:171-179.

16. Zhang J, Sun Y, Zhang H, Xu B, Zhang H, Song D: Preparation and application of triangular silver nanoplates/chitosan composite in surface Plasmon resonance biosensing. *Anal Chim Acta* 2013, 769:114-120.

17. Aliev FG, Correa-Duarte MA, Mamedov A, Ostrander JW, Giersig M, Liz-Marzán LM, Kotov NA: Layer-by-layer assembly of core-shell magnetite nanoparticles: effect of silica coating on interparticle interactions and magnetic properties. *Adv Mater* 1999, 11:1006-1010.

18. Wang Y, Biradar AV, Wang G, Sharma KK, Duncan CT, Rangan S, Asefa T: Controlled synthesis of water-dispersible faceted crystalline copper nanoparticles and their catalytic properties. *Chem Eur J* 2010, 16:10735-10743.

19. Liz-Marzán LM: Tailoring surface plasmons through the morphology and assembly of metal nanoparticles. *Langmuir* 2006, 22:32-41.

20. Liz-Marzán LM: Nanometals: formation and color. *Mater Today* 2004, 7:26-31.

21. Hoppe CE, Lazzari M, Pardiñas-Blanco I, López-Quintela MA: One-step synthesis of gold and silver hydrosols using poly (N-vinyl-2- pyrrolidone) as a reducing agent. *Langmuir* 2006, 22:7027-7034.

22. Sakai T, Alexandridis P: Mechanism of gold metal ion reduction, nanoparticle growth and size control in aqueous amphiphilic block copolymer solutions at ambient conditions. *J Phys Chem B* 2005, 109:7766-7777.

23. Sardar R, Park J, Shumaker-Parry JS: Polymer-induced synthesis of stable gold and silver nanoparticles and subsequent ligand exchange in water. *Langmuir* 2007, 23:11883-11889.

24. Pellegrino T, Kudera S, Liedl T, Javier AM, Manna L, Parak WJ: On the development of colloidal nanoparticles towards multifunctional structures and their possible use for biological applications. *Small* 2005, 1:48-63.

25. Boyer D, Tamarat P, Maali A, Lounis B, Orrit M: Photothermal imaging of nanometer-sized metal particles among scatterers. *Science* 2002, 297:1160-1163.

26. Hussain I, Graham S, Wang ZX, Tan B, Sherrington DC, Rannard SP, Cooper AI, Brust M:Size-controlled synthesis of near-monodisperse gold nanoparticles in the 1–4 nm range using polymeric stabilizers. *J Am Chem Soc* 2005, 127:16398-16399.

27. Wang Z, Tan B, Hussain I, Schaeffer N, Wyatt MF, Brust MJ, Cooper AI: Design of polymeric stabilizers for size-controlled synthesis of monodisperse gold nanoparticles in water. *Langmuir* 2006, 23:885-895.

28. Huber K, Witte T, Hollmann J, Keuker-Baumann S: Controlled formation of Ag nanoparticles by means of long-chain sodium polyacrylates in dilute solution. *J Am Chem Soc* 2007, 129:1089-1094.

29. Ershov BG, Henglein A: Reduction of Ag+ on polyacrylate chains in aqueous solution. *J Phys Chem B* 1998, 102(52):10663-10666.

30. Ershov BG, Henglein A: Time-resolved investigation of early processes in the reduction of Ag+ on polyacrylate in aqueous solution. *J Phys Chem B* 1998, 102:10667-10671.

31. Kiryukhin MV, Sergeev BM, Prusov AN, Sergeev VG: Photochemical reduction of silver cations in a polyelectrolyte matrix. *Polym Sci Ser B* 2000, 42:158-162.

32. Kiryukhin MV, Sergeev BM, Prusov AN, Sergeev VG: Formation of nonspherical silver nanoparticles by the photochemical reduction of silver cations in the presence of a partially decarboxylated poly(acrylic acid). *Polym Sci Ser B* 2000, 42:324-328.

33. Rivero PJ, Goicoechea J, Urrutia A, Arregui FJ: Effect of both protective and reducing agents in the synthesis of multicolor silver nanoparticles. *Nanoscale Res Lett* 2013, 8:1-9. PubMed Abstract | BioMed Central

34. Rivero PJ, Urrutia A, Goicoechea J, Rodríguez Y, Corres JM, Arregui FJ, Matías IR: An antibacterial submicron fiber mat with in situ synthesized silver nanoparticles. *J Appl Polym Sci* 2012, 126:1228-1235.

35. Rivero PJ, Urrutia A, Goicoechea J, Zamarreño CR, Arregui FJ, Matías IR: An antibacterial coating based on a polymer/sol- gel hybrid matrix loaded with silver nanoparticles. *Nanoscale Res Lett* 2011, 6:X1-X7.

36. Decher G: Fuzzy nanoassemblies: Toward layered polymeric multicomposites. *Science* 1997, 277:1232-1237.

37. Lee D, Cohen RE, Rubner MF: Antibacterial properties of Ag nanoparticle loaded multilayers and formation of magnetically directed antibacterial microparticles. *Langmuir* 2005, 21:9651-9659.

38. Wang TC, Rubner MF, Cohen RE: Polyelectrolyte multilayer nanoreactors for preparing silver nanoparticle composites: controlling metal concentration and nanoparticle size. *Langmuir* 2002, 18:3370-3375.

39. Logar M, Jančar B, Suvorov D, Kostanjšek R: In situ synthesis of Ag nanoparticles in polyelectrolyte multilayers. *Nanotechnology* 2007, 18:325601.

40. Gao S, Yuan D, Lü J, Cao R: In situ synthesis of Ag nanoparticles in aminocalix [4] arene multilayers. *J Colloid Inter Sci* 2010, 341:320-325.

41. Rivero PJ, Urrutia A, Goicoechea J, Matias IR, Arregui FJ: A Lossy Mode Resonance optical sensor using silver nanoparticles-loaded

films for monitoring human breathing. *Sens Actuators B* 2012. In press

42. Zan X, Su Z: Incorporation of nanoparticles into polyelectrolyte multilayers via counterion exchange and in situ reduction. *Langmuir* 2009, 25:12355-12360.

43. Yoo D, Shiratori SS, Rubner MF: Controlling bilayer composition and surface wettability of sequentially adsorbed multilayers of weak polyelectrolytes. *Macromolecules* 1998, 31:4309-4318.

44. Sergeev BM, Lopatina LI, Prusov AN, Sergeev GB: Borohydride reduction of AgNO3 in polyacrylate aqueous solutions: two-stage synthesis of "blue silver". *Colloid J* 2005, 67:213-216.

45. Sergeev BM, Lopatina LI, Prusov AN, Sergeev GB: Formation of silver clusters by borohydride reduction of AgNO3 in polyacrylate aqueous solutions. *Colloid J* 2005, 67:72-78.

46. Sergeev BM, Lopatina LI, Sergeev GB: The influence of Ag + ions on transformations of silver clusters in polyacrylate aqueous solutions. *Colloid J* 2006, 68:761-766.

47. Shiratori SS, Rubner MF: pH-dependent thickness behavior of sequentially adsorbed layers of weak polyelectrolytes. *Macromolecules* 2000, 33:4213-4219.

48. Choi J, Rubner MF: Influence of the degree of ionization on weak polyelectrolyte multilayer assembly. *Macromolecules* 2005, 38:116-124.

49. Urrutia A, Rivero PJ, Ruete L, Goicoechea J, Matías IR, Arregui FJ: Single-stage in situ synthesis of silver nanoparticles in antibacterial self-assembled overlays. *Colloid Polym Sci* 2012, 290:785-792.

50. Newman JDS, Blanchard GJ: Formation and encapsulation of gold nanoparticles using a polymeric amine reducing agent. *J Nanopart Res* 2006, 9:861-868.

Citations

CHAPTER 1

Katayoon Kalantari, Mansor B. Ahmad, Kamyar Shameli, Mohd Zobir Bin Hussein, Roshanak Khandanlou, and Hajar Khanehzaei, "Size-Controlled Synthesis of Fe_3O_4 Magnetic Nanoparticles in the Layers of Montmorillonite," Journal of Nanomaterials, vol. 2014, Article ID 739485, 9 pages, 2014. doi:10.1155/2014/739485.

CHAPTER 2

Han-Fu Hsu, Ping-Chen Tsai, and Kuo-Chang Lu, Single-Crystalline Chromium Silicide Nanowires and their Physical Properties, doi:10.1186/s11671-015-0776-8.

CHAPTER 3

Karolina Urbas, Malgorzata Aleksandrzak, Magdalena Jedrzejczak, Malgorzata Jedrzejczak, Rafal Rakoczy, Xuecheng Chen, and Ewa Mijowska, Chemical and Magnetic Functionalization of Graphene Oxide as a Route to Enhance its Biocompatibility, Nanoscale Res Lett.2014; 9(1): 656.Published online 2014 December 4.doi: 10.1186/1556-276X-9-656.

CHAPTER 4

Qi Zhu1, Zhixin Xu1, Ji-Guang Li12, Xiaodong Li1, Yang Qi3 and Xudong Sun1, Hydrothermal-assisted exfoliation of Y/Tb/Eu ternary layered rare-earth hydroxides into tens of micron-sized unilamellar nanosheets for highly oriented and color-tunable nano-phosphor films, doi:10.1186/s11671-015-0828-0.

CHAPTER 5

JitKang Lim, Swee Pin Yeap, Hui Xin Che, and Siew Chun Low, Characterization of Magnetic Nanoparticle by Dynamic Light Scattering, doi:10.1186/1556-276X-8-381.

CHAPTER 6

Sergey V Dorozhkin, Calcium Orthophosphate Coatings, Films and Layers, doi: 10.1186/2194-0517-1-1.

CHAPTER 7

Pedro Jose Rivero, Javier Goicoechea, Aitor Urrutia, Ignacio Raul Matias, and Francisco Javier Arregui, Multicolor Layer-by-Layer Films using Weak Polyelectrolyte Assisted Synthesis of Silver Nanoparticles, doi:10.1186/1556-276X-8-438.

Index